Guide to Source Inspection (Fixed Equipment)

Clifford Matthews
BSc, CEng, MBA, FIMechE

Edif NDE: A RINA Company, Leatherhead, UK

© Clifford Matthews 2017

This publication is in copyright. All rights reserved. Subject to statutory exception and to the provisions of relevant collective licensing agreements, no reproduction of any part may take place without the written permission of Edif ERA Ltd. Enquiries should be addressed to The Training Manager at Edif/ERA Ltd.

First published 2017

Typeset by Data Standards Ltd, Frome, Somerset, UK

Printed by Page Bros, Norwich, Norfolk, UK

A catalogue record for this publication is available from the British Library.

ISBN 978-0-9956762-0-6

Edif/ERA has no responsibility for the persistence or accuracy of URLs for external or third-party Internet websites referred to in this publication and does not guarantee that any content on such websites is, or will remain, accurate or appropriate.

The publishers and the author are not responsible for the reliability of any statement made in this publication. Data, discussions and any conclusions developed by the author are for information only and not intended for use without independent substantiating investigation on the part of potential users.

Contents

Acknowledgement	vi
Preface	vii
Part A Source Inspection of Fixed Equipment	**1**
Chapter 1 How to use this book	**3**
Chapter 2 Source inspection – what's it all about?	**6**
Chapter 3 The role of the source inspector	**10**
Chapter 4 The tactics of source inspection (how to do it)	**17**
4.1 Doing it well: the challenge you face	17
4.2 The tactics of NCRs	19
4.3 Rewind: the basic skills set	24
Chapter 5 Inspecting materials	**33**
5.1 Fitness for purpose (FFP) criteria for materials	33
5.2 Material forms	38
5.3 Common fixed equipment materials	40
5.4 Special materials	46
5.5 Material traceability	46
5.6 Test procedures and techniques	50
Chapter 6 Inspecting welding and fabrication	**64**
6.1 Welding inspection	64
6.2 Types of welds	68
6.3 Welding documentation	71
6.4 Test procedures and techniques	77
Chapter 7 Inspecting non-destructive examination (NDE)	**81**
7.1 Surface crack detection	81
7.2 Volumetric NDE	88
Chapter 8 Inspecting pressure vessels	**115**
8.1 Compliance and integrity criteria	115
8.2 Statutory certification	117
8.3 Working to pressure equipment codes	120
8.4 Inspection and test plans (ITPs)	125
8.5 Pressure testing	127

8.6 Visual and dimensional examination of vessels	133
8.7 Non-conformances and corrective actions	141

Chapter 9 Inspecting valves — 149
   9.1 ASME B16.34 — 149

Chapter 10 Inspecting painting — 155
   10.1 Paint types — 156
   10.2 Paint specifications and standards — 161
   10.3 Test procedures and techniques — 163

Chapter 11 Inspecting linings — 170
   11.1 Rubber linings — 170
   11.2 Specifications and standards — 173
   11.3 ITPs for lined equipment — 173
   11.4 Test procedures and techniques — 175
   11.5 Metallic linings — 180

Part B API SIFE Exam Preparation — 189

Chapter 12 The API Individual Certificate Programmes (ICP) — 191

Chapter 13 The API SIFE exam questions; what to expect — 195
   13.1 Exam question format: what to expect — 195
   13.2 Some hidden secrets about API exam questions — 195
   13.3 Final word: API exam questions and the three principles of whatever (the universal conundrum of randomness versus balance) — 203

Chapter 14 The SIFE body of knowledge (BoK) and study guide — 206
   14.1 The SIFE BoK: what's in it? — 206
   14.2 Dealing with such a large BoK (the road map analogy) — 206
   14.3 The SIFE study guide book — 209
   14.4 SIFE exam preparation — 225

Chapter 15 Metallurgy and materials: API RP 577 and related information — 238
   15.1 Introduction: metallurgy and materials – an important subject — 238
   15.2 The role of API RP 577 *Materials and metallurgy* — 239

Chapter 16 Non-destructive examination (NDE) – including ASME V and API 577 — 250
   16.1 Introduction: the importance of NDE knowledge — 250
   16.2 Next: the NDE sections of API RP 577 — 250

Chapter 17 Welding processes: API 577 and ASME IX — 258
   17.1 SIFE programme welding knowledge — 258
   17.2 A preliminary look at ASME IX — 259
   17.3 The ASME IX code rules covering the WPS, PQR and WPQ — 260

Chapter 18 Structural steelwork welding: AWS D1.1 — **270**
18.1 Introduction — 270
18.2 ASTM material 'general requirements' specification — 273

Chapter 19 Pipework ASME B31.3 — **277**
19.1 The SIFE BoK content — 277
19.2 ASME B31.3 quick reference points — 280

Chapter 20 Pressure vessels: ASME VIII-I — **288**
20.1 ASME VIII-I clauses — 288

Chapter 21 Valves and testing — **301**
21.1 SIFE BoK content — 301
21.2 Types of valve tests — 302

Chapter 22 Material verification (API RP 578) — **309**
22.1 The SIFE BoK content — 309
22.2 The content of API RP 578 — 310

Chapter 23 Surface preparation — **314**
23.1 The SSPC quality hierarchy — 314
23.2 Dry coating film thickness (DFT) measurement conformance SSPC paint standard no. 2 — 317

Appendix 1 Answers to Part B sample questions — **326**

Appendix 2 Steel terminology — **333**

Appendix 3 The ASME MDR form — **336**

Index — 340

Acknowledgement

Special thanks are due to Helen Hughes for her excellent work in typing the manuscript for this book.

Preface

The Edif group are pleased to publish this industry guide to the Source Inspection of Fixed Equipment (SIFE). The book is intended for inspectors or other individuals involved in the source inspection of new construction of pressure vessels, heat exchangers, tanks, fabricated piping and other fixed equipment.

Source inspection is a broad technical subject involving a wide variety of technologies and manufacturing capability. Large EPC contracts with multinational manufacture bring their own requirements of co-ordination, specification compliance and the quality of manufacture. Worldwide, the adoption of ASME, API and EN codes in many industries helps to bring a sound technical base to specifications and the activities of source inspection.

The technical competence of individuals is one of the main criteria of effective source inspection. Inspectors needs the ability to interpret and implement specification and code requirements across a range of fixed (static) and pressure equipment. Ongoing improvement and certification of source inspectors is a way to achieve this. This guide provides specific instruction for inspectors preparing for the API SIFE examination. This API SIFE programme is finding increasing acceptance in the inspection industry as a way of demonstrating inspector competence in commonly used codes and standards.

As a member of the Edif group of companies, Edif/NDE is well established in the source inspection field and is supportive of the API SIFE programme and examinations. Edif/NDE run an API SIFE exam training programme both for our own inspectors and clients, and to others, with the objective of helping to raise overall inspector competence in the industry.

For further details of the Edif/NDE training programmes visit our website www.edifgroup.com

Part A

SOURCE INSPECTION OF FIXED EQUIPMENT

Chapter 1

How to use this book

This book is intended to be a basic 'how to do it' guide to the inspection of fixed (i.e. static) equipment during manufacture. This is known in various industries as either *works* inspection, *shop* inspection or, more recently, *source inspection*.

Source inspection is a wide and multi-disciplinary subject as there are many different types of equipment around. This book therefore limits its scope to static equipment only – predominantly basic pressure equipment. It excludes fired equipment, such as boilers and some types of heat exchangers, as these are separate subjects in themselves.

You will not find this book to be a summary of detailed construction code information. There are hundreds of relevant codes from the petroleum/petrochemical industry alone – thousands if you take into consideration other industries. There are many code references spread around the book, however, where you will be referred to a relevant code or recommended practice (RP) document relevant to the source inspection activity. If you use these, then make sure to use the current edition or amendment.

Although the scope of static equipment covered in the book sticks to fairly basic categories (materials, pipework, vessels, valves and so on), these have been chosen to represent the most common items that you will meet as a source inspector (SI). Specialised items such as wellhead equipment, very-high-pressure vessels, bespoke research equipment, and so on, may have specialist material or functional testing requirements, but the basic skills and principles of source inspection are not *that* different.

Will reading this book teach me to be a source inspector?

Not entirely, but it is intended to give you a pretty good start. Source inspection requires a combination of code familiarity plus sound (and wide) engineering knowledge, sharpened and tempered (at the same

FIG 1.1
- In this book -
A and B summary boxes

PART A: SOURCE INSPECTION IN PRACTICE

JOB WELL DONE
- These boxes show you key points in doing your job properly

WATCH OUT FOR:
- These warn you of common NCRs (and things to watch for)

PART B: THE API SOURCE INSPECTOR FIXED EQUIPMENT (SIFE) EXAMINATION PROGRAMME

KEY LEARNING PRINCIPLES
- Principles (rather than technical facts) of the API (SIFE) programme

EXAM QUESTION ALERT!
- Common subjects for API (SIFE) examination questions
- Do not ignore these

time) by a healthy dose of experience. You also need to start off with the correct idea as to what the objectives of source inspection actually are, and the best way to achieve them. This then makes your experience *count*.

Why are there two Parts A and B to this book?

Part A covers the general objective and principles as set out above, accompanied by some general 'lowest common denominator' technical requirements. Part B addresses specifically the content of the API source inspector fixed equipment (SIFE) examination. This is a recent development by the American Petroleum Institute (API) to extend their Inspector Certificate Programme (ICP) into the field of source inspection. Part B of the book is specifically intended to be useful to candidates planning to attempt this examination.

Parts A and B of this book have different formats of summary boxes provided within their various chapters. Figure 1.1 shows the situation. Please concentrate carefully on these summary sheets – they are based on accumulated experience of performing source inspections across many countries and industries.

Chapter 2

Source inspection – what's it all about?

First – the name explained

Source inspection as we are now calling it, should perhaps be entitled *Source inspection and quality surveillance*. The word 'quality' can be a misleading one, unfortunately, calling up images of the writing of procedures and third-party quality audits to meet the requirements of ISO 9000 or some similar standard covering quality systems.

The justification for leaving the word 'quality' out of the title is that, although source inspection doubtless takes place within the boundaries and confines of a quality system or two, these systems are not put in place, or owned, by the source inspector (SI). The SI's role is more to provide a contribution to monitoring how a manufacturer's or contractor's quality system is working for a particular job. This is useful input. Formal 'quality audits' (there's that word again) are interesting as a bit of organisational theatre, but rarely reflect the reality of an individual manufacturing job or contract. Done properly, source inspection is a way of filling the gap, providing real feedback about just how good the shop-floor activities of a company actually are in following the client's codes, standards and specifications that they are supposed to be implementing.

Source inspection – why is it needed at all?

In an ideal world, it would not be necessary. Equipment would be manufactured precisely to specification, technical codes would be fully complied with, and manufacturers' and contractors' quality systems would control the whole thing, without the need for discussion, interruption or change. Some manufacturing processes do (almost) work like this. High-volume consumer goods, such as electrical goods and motor vehicle manufacture, are willing slaves to their own multiple embedded quality systems, ensuring a finished product of excellent reproducible quality. Source inspection by the final end-user therefore becomes obsolete, absorbed imperceptibly into the manufacturer's own system, and closed to outside view.

Unfortunately, larger, low-volume engineering items do not work like this. Pressure vessels, pipework systems and individually designed components just do not have the volume, the instant customer feedback, or the consistency of suppliers and sub-suppliers necessary to produce the reproducibility of the final output. Quality will vary both *between* suppliers and *within* the same supplier over time. When a non-compliance happens (as it will), the end-user will not find out until assembly on-site or, more likely, when the equipment develops a problem in use. By then it is too late – the non-compliance has crept through undetected and the damage has been done. This is why source inspection is required.

Contributory factors

Did you spot the change of terminology? *Poor quality* has been changed to *non-compliance*. Whichever of these terms you prefer (that is up to you), there is little doubt as to which will be the most useful to you. *Code compliance* is a definitive and provable concept. You can check a product for compliance with a client specification, material or construction (application) code and obtain a hard-and-fast answer 99% of the time. Most arguments based on interpretation and technical opinion will soon disappear with sufficient concentration on the documented technical requirements. The term *poor quality* is, of course, a different issue; along with its equally weak brother *poor workmanship*, it forms the perfect basis for long, multi-directional discussions with no real answer. If you do see the word *quality* again in this book, replace it in your mind with *compliance*, and you won't go far wrong.

There are five main reasons for non-compliances in multi-discipline engineering projects. Have a look at them now in Figure 2.1. Some are more topical than others but, on balance, they have not changed that much over time. Most get worse when a contract involves extensive subcontract manufacture and fabrication in different countries, as most do. Bring a lot of welding into the equation, and the chances of non-compliances start to multiply even further. Taken individually or, even worse, together, these five things are at the root of most of the non-compliance problems that, as an SI, it is your job to prevent.

FIG 2.1
The five main reasons for non-compliances

1. **Complex contract structures;** technical requirements get lost, misunderstood, or ignored in a weak contract structure comprising lots of different parties.

2. **Weak or mismatched specifications;** if client specifications conflict or compete with or do not recognise established construction codes (EN, ASME etc.), the result is confusion. Only optimists believe that confusion will naturally result in code compliance.

3. **Low-cost manufacture;** expect this trap to catch you for contract scopes involving low-value materials incorporated into highly labour-intensive shop fabrication. Cheap equipment is always cheap for a reason.

4. **Hurried timescales;** Doing things in a rush leads to corners being cut. Materials may be substituted, checks missed out, and a relaxed view taken on personnel suitability to do the job. Commercial pressures (they go together) make things worse in a weak attempt to justify the end result.

5. **Cultural differences;** Multinational manufacture brings together different cultures that take different views on compliance issues. What is a key technical compliance issue to one culture is interpreted as a loose, conveniently ignored irrelevance to another. Strangely, both think their view is correct.

Different types of source inspection

Source inspection is a wide subject used for all manner of different types of engineering equipment. Although the technical details may differ between equipment types, the nature of the source inspection activity itself remains much the same. The same objectives, principles, risks and pressures exist for most equipment. As an SI your skills are, in theory, transferable between disciplines, as long as you have the necessary technical knowledge to back them up.

Chapter 3

The role of the source inspector

The job

This is about *what* the job of the source inspector (SI) is, rather than *how* to achieve it (which comes in the next chapter). It's easy, the job is to be *successful*. Success means finding all the compliance-threatening issues that are there. Once you find them, you then become a key part of the mechanism for reporting and correcting them. All of this takes place in the manufacturer's works to ensure the problems never make it to the stages of site assembly and operation. Operational problems are therefore happily prevented before they appear. Sadly, this means you can expect little credit – your reward lies in the self-satisfaction achieved by doing a difficult job properly.

Success and comfort

Don't confuse success with *comfort*. Success at source inspection will almost certainly involve operating outside your own technical comfort zone. Inspection is about monitoring the work and performance of others. Most of these others will have daily involvement in their own technical subjects, and so will know more about it than you do. Inspection then involves finding problems, criticising or influencing the work of others. Not everyone will see this as being either useful or constructive – hence its position outside your comfort zone.

The management part

Although seen as a technical role rather than a management one, source inspection involves one of two bits of management. The first is easily recognisable to most SIs – the task of managing the parties 'down the chain', that is, contractors, manufacturers and sub-manufacturers. They are the ones who do the work and have to be influenced to get it right.

The second role is equally important, but not quite so obvious – the

FIG 3.1
Source inspector: The management role

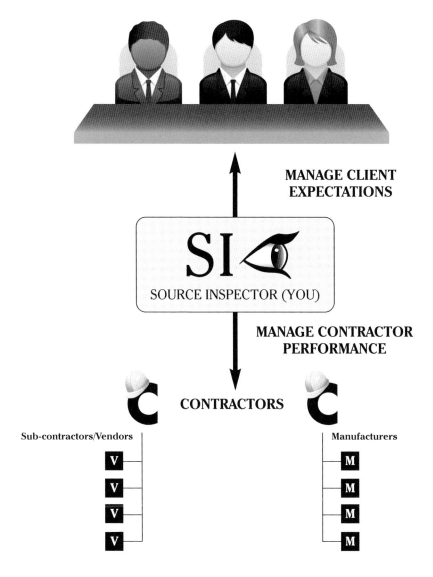

need to manage the expectation of *your client*. As a SI your client can be the equipment end-user, engineering procurement contractor (EPC) or third-party inspectorate body, depending on the specific contract details. Clients generally know *broadly* what they want in equipment performance terms, but their high-level view means that they cannot assimilate all of the technical detail. Figure 3.1 shows these two roles. Here are the results.

As a SI, you can expect your client occasionally to

- **place poor value-judgements** on which part(s) of a technical specification are important, and which are less so
- **misunderstand** the real objective of design/construction codes and their relationship to the fitness for purpose of an engineered item
- **consciously allow** commercial contract pressure (time or price) to override technical issues – kicking them down the road in the hope they will go away
- **be influenced** by their previous experience (good or bad) with individual manufacturers and contractors
- **mislead themselves** and concentrate on trivia.

In most large contracts, most of these issues will be there, to a greater or lesser degree, if you actively look for them.

The technical part

Within the general background of keeping client expectation and the activities of the manufacturers under control, the role of the SI is undeniably a technical one. Duties consist of

- **participating** in the steps of the inspection and test plan (ITP), namely, quality surveillance
- **witnessing** tests (of many types)
- **accepting** or rejecting test results against purchase specification and code requirements
- **compliance** (and non-compliance) reporting and final acceptance of manufactured equipment.

These four elements fit together to make up the overall picture of *compliance* of the equipment or system against its specified requirements.

Figure 3.2 shows the situation. Note how all of these activities are linked together by the ITP. This is most commonly prepared by the contractor/manufacturer and approved by all of the other parties

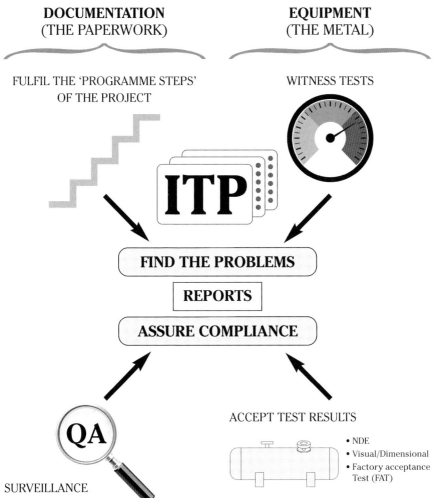

FIG 3.2
Source inspector: The technical role

ALL OF THESE ACTIVITIES ARE DOCUMENTED BY THE PROJECT INSPECTION AND TEST PLAN (ITP)

involved. It provides the central procedure document for the inspection-related activities of the manufacturing stages of the equipment. For ASME code equipment, full completion and acceptance by the necessary parties of the ITP culminates in the issue of the ASME-code-specified manufacturer's data report (MDR).

Do source inspectors inspect equipment or paperwork?

The answer to this lies in the *way* that a SI chooses to do their job (the subject of the next chapter). From a purely objective viewpoint, the answer is easy – it is the equipment itself that performs the function that the client is purchasing. No stack of drawings, computer files or test certificates has ever transported fluid or acted as a pressure source in an engineering process.

In reality, the physical inspection of engineering items is always supported by a review of the paperwork that accompanies it. It helps to form a record of what was done, for review by the various parties concerned. Be careful not to think of it as a *replacement* for the physical activities (there it is on the left-hand side of Figure 3.2); it is only part of the story.

Source inspectors witness tests

The technical role of the SI includes witnessing tests. There are many types, but common ones are

- material tests, for mechanical properties of parent material or welds
- non-destructive tests/examinations (NDE)
- pressure tests
- functional tests, on valves, for example.

Tests normally follow some kind of written procedure to make sure they are done properly. Some tests are just simple visual and dimensional checks against weld maps, weld specifications or construction/general arrangement (GA) drawings. Others are more complicated, requiring specialist test equipment in the laboratory or workshop.

Acceptance criteria

The physical act of being present to witness a test or performing a visual/dimensional inspection is not difficult; the decisions about the output are the hard part. Every test and inspection has acceptance criteria of some sort. In some cases it is as simple as compliance with the

FIG 3.3
Job well done: Source inspection

The role of the source inspector (SI) is to

Give priority to the physical items of equipment (the metal).

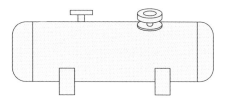

AND NOT TO

Rely entirely on documents, certificates and test reports (the paperwork).

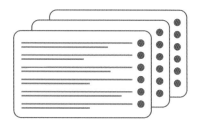

JOB WELL DONE

dimensions shown on an engineering drawing. More often, however, the situation will be more complicated, with specification or (more commonly) code-specified acceptance criteria that have to be met. Material properties, NDE and pressure tests are all good examples of these.

Some acceptance criteria can be shown directly on an ITP, but more often are cross-referenced in the ITP back to a published design/construction code. We will look at the individual content of these in future chapters. For the moment, recognise the fact that understanding and judging against acceptance criteria is one of the key roles of the SI.

Chapter 4

The tactics of source inspection (how to do it)

In the previous chapter we looked at *what* the source inspector (SI) does. This chapter is about *how* to do it. As before, we shall start from the premise that *success* is defined as doing it *well*.

4.1 Doing it well: the challenge you face

Making a good job of source inspection is about combining a high level of technical knowledge with an understanding of how the activity works, with its characteristics, challenges and pitfalls. Figure 4.1 shows the size of the task. The first step in finding a way through is to define clearly what the objective of the source inspector (SI) is. We know this from the previous chapter – it is to find the issues of technical non-compliance that have been produced during the equipment manufacturing process. The job of the SI therefore starts with finding *problems*.

Do you believe that problems are there?

To be a successful SI, you need to start out by believing that problems (let's call them non-conformances (NCRs)) are there to be found. They can appear anywhere: incorrect materials, wrong welding procedures (WPSs), non-compliances with drawings, missed tests, incorrect test procedures, out-of-code test results, the list is endless. Some will end up never affecting significantly the integrity or functional performance of the item, but hidden among them are those that will.

Now we have an important principle of source inspection, which is

- always assume your manufacturing project is *packed full* of NCRs

and

FIG 4.1
Source inspection:
The challenges you face

The wide scope of specifications, standards, recommended practices and published documents involved

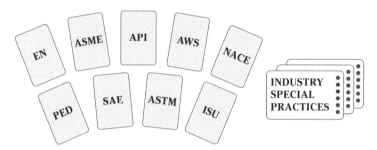

Most SI information is self-taught

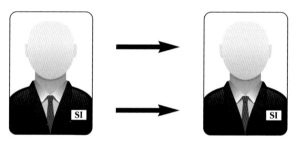

Inspection is about being outside your comfort zone (for much of the time)

- if you haven't found any, it is probably because you are not trying hard enough, or are not sure what you are looking for.

In real manufacturing of static equipment these statements hold true for such a high percentage of the time (75–80% *at least*), that together they form a wise premise to follow.

What if you look for compliances rather than non-compliances?

If you start off by looking for compliances, things that are correct, you will find them without difficulty. Everywhere you look, you will find evidence of the status quo, telling you the story that everything is fine. The more you look, the more you will find, reinforcing your rapidly increasing viewpoint that no intervention is required by you. This will feel good because it is a natural tendency for everyone to look for their comfort zone. It is all illusion of course; you merely looked with the objective of justifying to yourself the predetermined answer that you wanted. Do this and the non-compliances will escape you. This is the danger.

Are you the solver of problems you find?

Not always. Identifying non-compliances is the first step in getting them solved, but correcting them may involve a process that is not fully within your remit or ability as a SI. Welding issues, material substitutions, design changes and repairs are good examples of this.

For simple non-compliances, many can actually be closed out 'at source' before they ever attain the status of an NCR. It is your job as an SI to do this, to prevent NCRs about easily solvable points taking on a life of their own when this is not required. Once released into the project structure, NCRs have an awful habit of reverberating around, becoming progressively exaggerated and misinterpreted, sucking in money and time resources as they go. Sometimes they are conveniently dropped or kicked forward to be 'corrected on site', where they are rapidly swamped by other priorities. Solving NCRs at source is *always better* if you can.

4.2 The tactics of NCRs

Figure 4.2 shows the tactics of NCRs. This four-step approach works well in most situations. The four steps do not stand alone, however, they need to be combined with the basic skills set shown later in Figure 4.3.

FIG 4.2
The tactics of NCRs

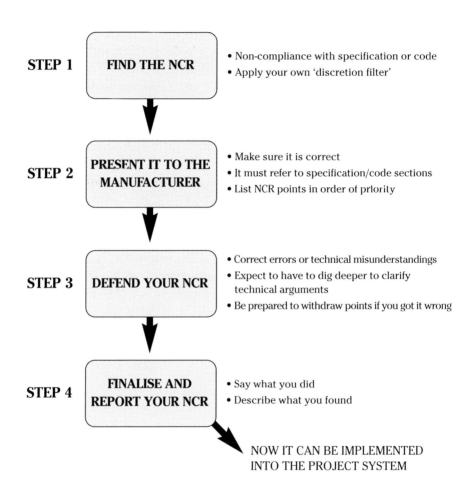

AND FOR THE SKILL SET THAT YOU NEED; SEE FIG 4.3

Let's look at the steps shown in Figure 4.2 one at a time, in the order in which you should use them.

You should only issue a NCR against a specific, clearly documented requirement of a client specification or technical code cited in the contract. Contract documents are structured into a hierarchy, with some clauses taking precedent over others, so you cannot rely too heavily on those with little contractual muscle. Inspection and test plans (ITPs) are a good example of a document that is normally well down in the hierarchy (see Figure 4.4 later). Don't be misled into making a big issue about the following points either

- cosmetic painting details (chips, colour shades and so on)
- items that are within specification or code but which you personally think are poor quality or poor workmanship
- the hundreds of minor paperwork and administrative imperfections that are waiting to waste everyone's time if you raise them as NCRs.

'Missing paperwork' NCRs

You have to be careful about raising NCRs where the *only* issue is missing paperwork. This applies to almost any of the paperwork procedures, reports, test results or certificates shown on the ITP as being required. This part of the SI role is easy to get wrong – it's a delicate balancing act of decisions, which, if taken wrongly, will cause you either to miss a vital non-compliance point *or* waste everyone's time within a spiral of NCR activity that will ultimately lead nowhere. Consider the following points.

- Just because a paper record of a test or activity is not present, does not mean it does not exist.
- A missing paper record of a test does not necessarily mean the test has not been done.
- If you raise a NCR based solely on missing paperwork, it will, 99% of the time, miraculously appear later, thereby closing out your NCR (so it has effectively led nowhere).

List your NCRs in order

The best way is simply to list your NCRs in the order in which the manufacturing process progresses, that is, starting with materials, progressing into the stages of fabrication and NDE, through to final testing and documentation. Following the chronological steps of the ITP is the easiest way to do this – it keeps everything in a logical order.

Finding and listing NCRs takes time, because it is essential that you find them all, and that they are correct. The following points are key.

- Make your NCRs precise, containing statements of technical fact.
- Use correct technical terminology – do not waffle or make meaningless statements of general discontent.
- Link every one to the specification or code clause that you are saying has not been properly complied with. If you cannot quote one, you should not be issuing an NCR on that subject.

Apply your 'discretion filter' on minor points

The purpose of an NCR is to raise awareness and stimulate corrective action on non-compliance points that *matter*. You won't improve your reputation as an SI by concentrating on minor or insignificant items of paperwork. The power of this point (it is not your power, unfortunately) is that you are forced to be satisfied once the paper appears. This is because you raised the NCR based *solely on the existence of the paperwork, not on the activity or test to which the paperwork relates*. This is the weakness of paperwork-only-based NCRs – real mechanical integrity issues can be hidden or missed, obscured by the focus on, and shuffling of, all that paperwork.

Presenting the NCRs to the manufacturer

First rule:

> You must present the NCRs when you are *in the works*, giving the manufacturer the opportunity to respond to it, there and then. If you issue it later when you are back in the office (because you were not sure, or you are frightened of disagreement or technical questions), your technical creditability will sink to near zero. Word will travel quickly, but no-one will tell you, so you'll feel fine, happily.

Second rule:

> Be prepared to defend using technical argument everything you have written.

Defending your NCRs

It is perfectly normal for a manufacturer to be in disagreement with at least some of the points you present in your NCRs. Most disagreement is about the relevance or interpretation of code or specification clauses rather than technical facts. Expect differences of opinion on

- the relevance of code/specification clauses to the job in hand
- interpretation of code clauses themselves
- the meaning of statements such as *equivalent materials*, *relevant acceptance criteria*, *reasonably practicable* and similar.

In practice, defending your NCR points is a process of technical clarification, investigating and discussing each technical point in turn until things become clearer. As the whole process is founded in technical fact, the real situation will reveal itself quicker than you think, as long as both sides put sufficient technical effort into looking up reference documents, code clauses, or whatever, and keep their focus on the points under discussion (see Figure 4.3 and subsequent discussion of *focus*).

Don't make the mistake of confusing NCR discussions with some kind of delicate negotiation procedure. There is nothing to negotiate and nothing to trade off. It is simply a search for the *technical truth*. Getting this stage right is one of the core skills of source inspection. At first you will find it difficult – your technical knowledge (or lack of it) will be laid bare in the face of technical opposition from those who know more about the specific contract or equipment than you do.

With experience, you should recognise that NCR clarification activities split into two sequential stages. The *first stage* is where the most easily clarified points of the NCR are decided and your initial non-compliance points are each agreed with or disagreed with. Both constitute *success* at this stage, because the situation has been made clear. After a delay in which documents and other technical people are consulted, along comes the *second stage* – a more detailed discussion of those parts of disagreement left over from stage one. Just when you thought it was finished, up pop again the technical discussions on code interpretation, NDE acceptance criteria, performance test results and so on. This time the arguments become more technical, requiring you to dig deeper into your knowledge to support the technical opinions you have expressed in your NCR.

During these clarification stages, you must be prepared to withdraw any NCR points that are proved to be incorrect. There is no need to be concerned about this – it is a perfectly normal situation, given the number and complexity of technical points that arise. As long as you let the NCR process develop slowly, and do not rush to instant conclusions, this will just be seen as an unremarkable part of the process and your credibility will remain intact. The opposite is also horribly true – if you continually try to defend a technically incorrect position, then your credibility will soon drift away. This situation is often seen.

Finalise and report your NCR

Once all the hard clarification work and technical argument is done, the next stage is easy. It is just the administrative action of distributing the NCR into whatever contract system you are working under. Copies will go to various people, some more involved than others. Do not forget the following important points.

- Get the manufacturer to sign and accept a copy of your NCR. Suggest they explain in the NCR any points that they do not accept.
- Check you have enclosed enough information that it can be interpreted by anyone who may need to do so, but was not present at the inspection (designer, metallurgist and so on). This will help in the resolution of any outstanding points of technical disagreement.
- Once you have finalised it, you cannot easily change your mind (unless new technical information comes to light), so make sure it is *correct*.

4.3 Rewind: the basic skills set

Now that we have looked at the tactics of source inspection, and in particular the difficult business of raising and clarifying NCRs, you can see the extent of the challenge if you want to do it well. Success relies on you having a solid depth of technical knowledge on the correct subjects, tempered by a mix of forward thinking, suspicion, curiosity, intuition and experience. Taken together, these can be shaken into a set of five basic skills. There are others, no doubt, but these five will give you most of what you need. Look at them now in Figure 4.3. We shall look at them separately, but remember that they all work together (with fuzzy boundaries) during the activities of a source inspection.

Focus

Focus means keeping sight of priorities. It is more than likely, in fact almost certain, that the various parties present at a source inspection will each have a different focus on events, even though their technical objectives are broadly the same. This gives a real potential for

FIG 4.3
Source inspector: The basic skills set

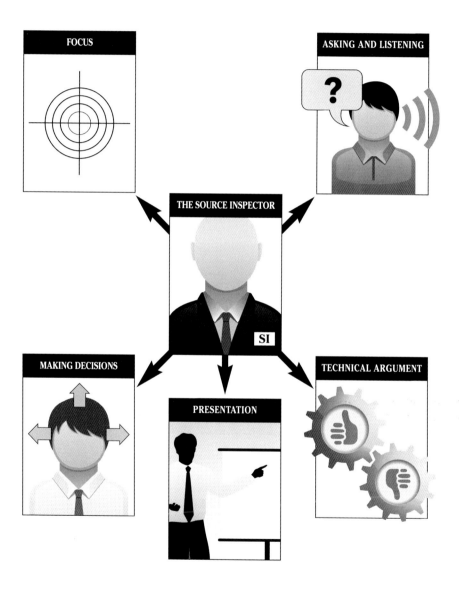

misdirection, time delays and extra costs, before a consensus agreement or solution can be found. Good source inspection involves manging these different foci.

Look at how it works in practice. During a source inspection on, for example, a pressure vessel, the main thread of evaluating code compliance and integrity is complicated by the development of many interesting, mainly technical, diversions, as the (up to five or six) parties present interact. These diversions start off slowly, but issues of material traceability, testing techniques, workmanship, painting and packing, weld specifications and other things soon develop as side issues of the main theme. Commercial issues and questions of interpretation also appear to further complicate the arguments. Under such conditions the side issues can become very effective at blocking your focus on integrity and code compliance because side issues are often more interesting, and easier to discuss, than the main issues. They can be an easy way out, so you must guard against them. Fortunately the rules are simple.

- Get a clear focus on integrity and code compliance.

then

- Just keep on coming back to it – again and again.

This is a loop, and it is a good idea to make it a 10-min loop. This means that you can play a full part in side issue discussions, but every 10 min you need to bring the subject of the discussion back round to your integrity and specification compliance focus. Any longer than 10 min and the side issues may well grow in priority until you are seen as unreasonable in trying to change the subject. Less than 10 min and you may miss an important point that the side discussions reveal; so 10 min is about right. Figure 4.4 shows a typical example of a focus loop that can occur in practice.

Asking and listening

The moderate view is that the activities of source inspection are those of an inquiry, rather than a trial. All parties start out neither innocent nor guilty and will remain that way, whatever the outcome of the inspection. Accepted. It is also a fact, however, that source inspection is about questioning the actions and technical knowledge of competent contractors and manufacturers and can therefore be a stressful activity for all parties. Perhaps the best general guidance to the SI is to try to project a modest attitude of searching for the truth, rather than

FIG 4.4
Source inspection: Keeping your focus

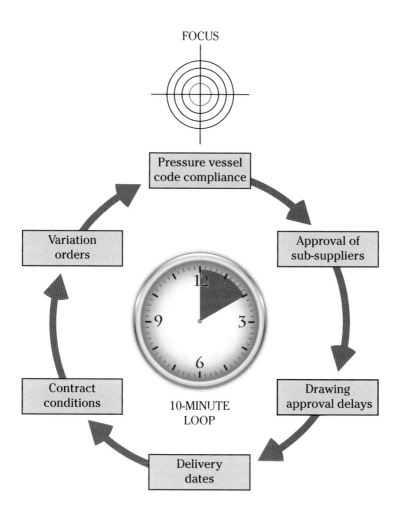

attempting to appear too confident, or authoritative, or even apologetic. You will see some SIs who almost apologise for being in the works. Granted, there are many different ways to do it, but over time the honest truth-seeking approach, albeit hard to maintain, probably works best.

Armed with this, you can start asking questions. A good technique to master, which will serve you in all kinds of inspection situations, is that of *chain questioning*. This is a very effective way of getting at the truth. The technique involves spending most of your time (upwards of 80% of the time you spend speaking) asking questions; not in a suspicious or confrontational manner (although these may have their place), but asking, just the same. The precision of the questions is important – they need to have two main properties. First, they should be capable of being answered by the manufacturer or contractor in a way that allows *verification*, that is, the manufacturer can prove to you that what he says is true – he can provide evidence. This means you must be accurate in what you ask. Second, the questions must be *chained* – each one following on from the last answer, developing progressively better levels of resolution on a connected subject in question, rather than hopping about from subject to subject. Note that it is essential that you obtain verification of the previous answer before developing the next question in the chain, otherwise future questions and answers can become increasingly hypothetical and the integrity of the chain breaks down. In most situations, a mixture of closed and open-ended questions seems to work best. Done correctly, you will find that a well-chosen chain of four or five questions with tenacious perusal of verification of each answer will strip the veneer off most source inspection situations.

A final point on questioning. Remember that source inspection is an *inquiry*. Once you have asked a question, it is not wise to answer it yourself. Don't even start to. Just wait. Thoughtful silence will normally bring the answer.

Making decisions

Making decisions is what source inspection is all about. Almost every inspection visit will involve decision making about whether to accept test results, compliance with acceptance criteria or documentary evidence of this or that. The interesting point about decision making in a SI context is that 90% of the effort is used in *obtaining the technical information* to enable you to make the decision, so the final act of making the decision is the easy part.

The fact that can make source inspection uncomfortable for some people is that most decisions will need to be made by *you,* yourself. Even if you are working within a large company structure employing specialists in design, welding, materials and so on, it is wise to expect that not much specialist advice will be available to you down at the level of a shop-floor inspection in a manufacturer's works, well away from your office. This is more a function of practicality than a conscious desire of your employer to leave you unsupported. You have to be prepared therefore to take inspection decisions yourself.

Not everyone likes making decisions. You may be worried about your lack of knowledge, or being overruled or criticised by the manufacturers, your client or even your own employer, for making a decision that not everyone likes. Unfortunately, this is all part of the SI role; making decisions is what you are there for. Have a look at the following points.

- Making decisions **in the works**, yourself, is the most efficient way of doing it. If you defer this role to absent specialists it will take much longer and cost a lot more, with no guarantee of a more correct result.
- If you **don't make** decisions during a source inspection you will lose credibility with the manufacturer (first) and then your client (later).
- It is your job. SIs are paid to make decisions about compliance. That is what you are there for.
- On balance, an **incorrect** decision is better than **no** decision. It is easier to resolve.

Making decisions, particularly correct decisions, becomes easier the more source inspection experience you get. You will also improve with practice, as you progressively discover which decisions work and which do not. Every time you consciously avoid making a decision you decide to restrict your own learning. It is worth repeating, once again, that as an SI your role is to *make decisions* not avoid them, or wait for someone else to make them for you.

Technical argument

Having previously discovered that 90% of the effort involved in decision making is not actually involved in making the decision, you should not be surprised to learn that technical argument does not actually contain much *argument*. Instead, think of technical argument about any requirement of an ITP as being more of a *technical discovery* process. The argument does three things.

- *Identifies* the technical issues that warrant discussion, eliminating those that do not.
- *Encourages* focus on the ITP, client specification and applicable code clauses that really matter, rather than wasting time with subjective viewpoints.
- *Stimulates* your learning.

From an on-going perspective, the last point is the most important to you, the SI. The sheer variety of equipment functions, types and manufacturing practices that make up the world of the SI will always mean that you have lots to learn. Even 30 years of source inspection experience will not fill in all of the gaps. This means that you need to find some way to accelerate your learning speed above the 'real-time' knowledge you will absorb by participation in repetitive source inspections.

Technical argument will do this for you. You will learn three times as quickly, with more depth, when finding the answer to a point of technical dispute than you will if all you discuss is consensus. Disagreement means you have to dig deeper to get the correct information to make your point. The other parties will do the same, raising the technical level of the whole discussion. High-quality technical fact will replace subjective viewpoint based on half-remembered opinion or hearsay and, as the stakes rise, so does the quality of your knowledge.

What if you lose a technical argument? This is even better. Once you have made a technical decision, then being proved (fairly) wrong is just about the best learning experience you can get. You can see now the debilitating effect of *not* making decisions – you will learn at a pedestrian rate, if at all.

Butterfly thinking: the dangers of multi-tasking

By design, your brain is very poor at multi-tasking. In half a million years or so of human development most inventions and innovations have come about as the result of an individual or group concentrating on a single thing, to the exclusion of, at the time, almost everything else. Continually jumping from task to task: documents, laptop, material code, telephone, ITP, phone, text, emails, shop-floor visit, phone (again), is just butterfly thinking and *does not fit in* with the way that you are

FIG 4.5
Job well done
Source inspection and how to do it

How did your last source inspection go?

Did you:

	YES	NO
• Go well prepared with copies of all the codes and specifications you need	☐	☐
• Anticipate what the real technical issues were?	☐	☐
• Find any* significant non-compliances	☐	☐
• Sucessfully argue all your NCR points	☐	☐
• Make all the necessary decisions, rather than sit on the fence?	☐	☐

*If you didn't, were you really trying to or were you being a butterfly?

designed to solve problems and make decisions. This deficiency is hard-wired into you, so you cannot change it. So, it is worth remembering that

- multi-tasking produces poor decisions
- butterfly thinking works fine (for butterflies).

Final word – source inspector

As with all roles involving checking, inspecting and adjudicating on the work of others, you cannot expect source inspections to go smoothly all the time. If they do, reflect on whether you are putting sufficient effort into really finding the technical non-compliances that are there, or whether you have just become happy with the easy life of not finding any problems that may embarrass others or (more probably) yourself.

Either way, your success at source inspection is a close function of the breadth and depth of technical knowledge that you have in your head. In the remainder of Part A of this book we shall now look at the basic areas of technical knowledge that you need for the inspection of fixed equipment.

Chapter 5

Inspecting materials

Source inspection involves *verification* of the engineering materials used to construct equipment. It is not, strictly, about material *selection*, although this does come into the equation. Materials are one of the lowest common denominators of engineering practice, with the common steels and non-ferrous alloys used in many and varied types of fixed equipment. There are also many specialised materials, each with its individual niche of application and designed specifically to resist a particular regime of stress, temperature or process conditions.

Materials inspection is not just technical. Material technology is heavily standards-based, and the issue of material *traceability* occupies a prominent position. This is a procedural aspect of engineering manufacture that forms a common feature of the source inspection role, particularly in relation to pressure parts and structural components.

5.1 Fitness for purpose (FFP) criteria for materials

There are many different tests and analyses that are carried out on engineering materials. The purpose of almost all such tests is to verify one or all of three main FFP criteria: mechanical properties, temperature capability and positive identification.

Mechanical properties

The mechanical properties of a material are matched to its application by the equipment's designer, assisted by the accumulated experience of proven and accepted technical standards. The common mechanical properties are *tensile strength* (this is truly a 'strength' measurement), *ductility* (measured by percentage elongation and percentage reduction of area), *toughness* or resistance to shock (measured by impact value) and *hardness* (the ability to withstand surface indentation). You will see these parameters specified in material standards. Less commonly specified properties include fatigue and creep resistance, and resistance to corrosion. These properties can be considered as compound functions of the other mechanical properties, rather than primary properties in themselves. Figures 5.1–5.3 show the details.

The mechanical properties of an engineering material only have meaning as a set, rather than individually. Tensile strength or hardness, taken alone, is inadequate to describe fully the way in which a material performs; measures of ductility and toughness must be added to complete the picture. It is better therefore to think of this compliance criterion as consisting of compliance with a *set* of mechanical properties, even if one of the properties does not appear to be predominant when you consider the application of the particular material.

As a source inspector (SI) you will not normally experience a situation where you are obliged to select a material for a specific design application. The nearest you should come to this is when you have to interpret the material requirements written in a contract specification. This means that you need to develop your background knowledge of engineering materials; it is not always sufficient simply to check a set of test results against the listed values in a technical standard.

Temperature capability

The resistance of a material to elevated temperature has some relationship to its mechanical properties. Both European and American standards differentiate clearly between low-, intermediate- and high-temperature ferrous alloys. High temperature often means high pressure and the existence of unpredictable failure mechanisms such as creep. Source inspection is a wide, cross-disciplinary, technical subject, so temperature capability is one useful point of reference that will help you to keep things in perspective.

Positive identification

It is almost impossible to identify most materials by visual observation alone. For this reason positive identification is an important FFP criterion. It is not strictly technical – it has a procedural basis – but the activities form a well-defined element of engineering manufacture. We shall look at the detailed activities of material identification and the traceability chain later in this chapter.

Inspecting materials

FIG 5.1
The five main mechanical properties of metals
– visual reminders –

Strength

Tensile test piece

 Force

A *strong* material takes a larger stress before it yields

Ductility

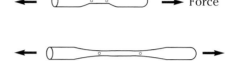 Force

A *ductile* material stretches and thins a lot before breaking

Malleability

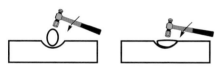

A *malleable* material can be formed by compression

Hardness

A *hard* material resists surface indentation and scratching

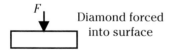 Diamond forced into surface

A *soft* material indents more easily under the same force

Toughness

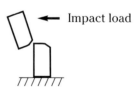 'Charpy' test piece

A *brittle* material cracks and breaks easily under impact loading

Impact load

A *tough* material deforms rather than breaks in a brittle manner

FIG 5.2
Mechanical properties of metals

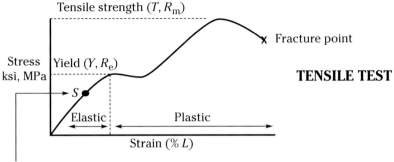

TENSILE TEST

Allowable stress (S) is in the elastic region (typically 66% Y or 25% T) or as defined by the application code, e.g. ASME VIII, B31.3 etc.

Specimen before tensile test   **Specimen after tensile test**

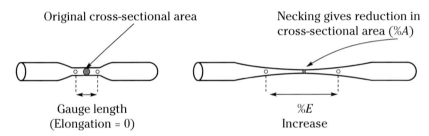

HARDNESS TEST

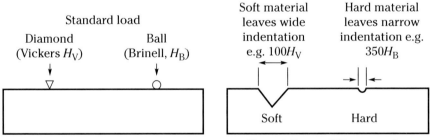

H_V and H_B are different scales
(but not too far apart)

FIG 5.3
What causes a metal's mechanical properties?

Mechanical properties are the result of a combination of four sets of factors

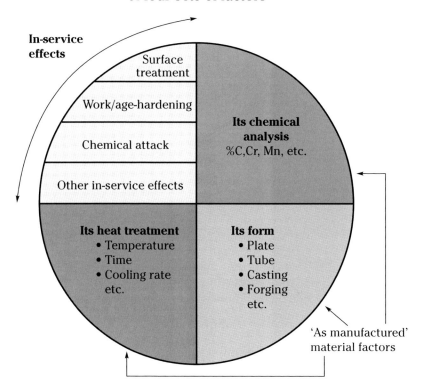

'As manufactured' material factors

Taken together, these produce the five main mechanical properties of metals

- Strength
- Ductility
- Malleability
- Toughness
- Hardness

5.2 Material forms

There is a finite amount of technical information that you need to know about materials of construction because technical issues regarding the common engineering materials found in a process plant are generally repetitive. A clear framework of the type of information that is useful, coupled with some experience of the issues in practice, will give you what you need to carry out effective source inspections in this area.

There are four predominant forms of material manufacture: **forgings, castings, plates (or sheets) and tubes** (see Figure 5.4). There are similarities in common technical areas such as material analysis and mechanical tests within each of the categories. The categories also reflect the way in which you will meet material documentation in a source inspection situation.

Acceptance guarantees

For the majority of the materials of construction included in a plant contract, there is no specific *acceptance guarantee*, as such. Instead there will be a set of material standards raised directly, or inferred, by the contract specification and its attendant document hierarchy. The requirements of these standards effectively form the acceptance criteria for the materials of construction. Practically, the scope and clarity of material standards are very good. Occasionally (perhaps 10% of the time) you will find special and overriding material performance requirements stated in the contract specification. These may take the form of explicitly stated values for chemical analysis or mechanical properties *or* references to more specialised, less well-known standards. You will see industry-specific standards such as the NACE (National Association of Corrosion Engineers) series referenced like this. Some of the more common special requirements are listed below.

- *Corrosion tests*. These are specialised tests to determine resistance to intercrystalline corrosion of austenitic stainless steel.
- *Compound distribution tests*. The most common one is the sulfur print, a chemical process used to obtain a visual print of the distribution of sulfur compounds in a material specimen.
- *FATT (fracture appearance transition temperature) test*. This is an extension of the impact test to determine the relative amounts of brittle and ductile failure areas on a fractured specimen over a range of temperatures.

FIG 5.4
Some common 'material form' steel standards

e.g. cold-formed pipe fitting

5.3 Common fixed equipment materials

Most fixed equipment contracts rely on well-known published material specifications. The acceptance criteria are well defined, with the level of definition being better in published material standards than in many equipment standards. This is generally because the technical scope is tighter for materials than for equipment. Material standards are configured in a particular way, based on the results of many years development. American and European standards that you will need follow similar structure and principles. Figures 5.5–5.7 show some commonly used types. Note also Figure 5.8 and Figure 5.9, which show the all-important role of carbon content. Appendix 2 gives further information on steel terminology.

Note the following points.

- Obtain access to the main reference documents for Euronorm (EN) and ASME materials. As a back-up be aware of the German reference document *Stahlschussel*. This lists and identifies the majority of ferrous materials manufactured throughout the world. It also shows equivalence between material designations (which can sometimes be confusing).
- Use the cross-references. If you look *inside* the front or back page of a material standard you will find a list of cross-referenced standards. Because of the way that standards have developed and spread over the years it is essential to check these standards to see how they impinge upon your compliance verification. Check the nearest ones carefully, watching for special requirements relating specifically to the four main material manufacturing processes (forging, casting, sheet and tube) mentioned earlier.
- Do not ignore superscripts, subscripts and footnotes. They are a feature of material standards and can involve quite important changes in the 'sense' of the information – particularly chemical analysis and heat treatment. Be careful also about the notes concerning the way in which acceptable mechanical properties are allowed to vary depending on the size of section or thickness of the material – there can be significant differences.
- Remember that source works inspection is about material *verification*, not material selection. Don't get too involved.

FIG 5.5
Typical pressure vessel plate: ASTM SA-516 Gr 60

ASTM SA-516 GR 60 is a common pressure vessel steel ('boiler plate')

Equivalent to
- BS1501-161-430A
- EN10028-P265-GH
- DIN 17155 HII

For thicknesses <40 cm it is supplied 'as rolled'

For thicknesses >40 cm it is supplied *normalised* to improve its impact strength (toughness)

Gr 60 denotes a minimum specified yield strength of 60 ksi

Chemical analysis

- Carbon (C) 0.18% max
- Manganese (Mn) 0.95–1.5%
- Copper (Cu) 0.3% max
- Molybdenum (Mo) 0.08% max
- Silica (Si) 0.4% max
- Nickel (Ni) 0.3% max
- Phosphorous (P) 0.015% max
- Niobium (Nb) 0.01% max
- Sulfur (S) 0.008% max
- Titanium (Ti) 0.03% max
- Aluminium (Al) >0.02%
- Chromium (Cr) 0.3% max

Note the cocktail of these trace elements added. Ni and Mn increase strength, Cr increases hardness

Mechanical properties

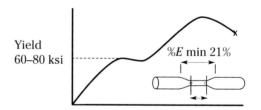

Yield 60–80 ksi

$\%E$ min 21%

FIG 5.6
A typical cast steel: ASTM A-216

Note the three grades available

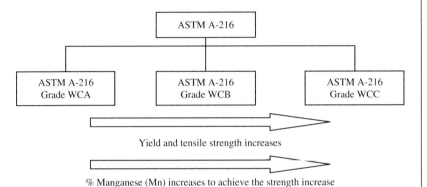

Yield and tensile strength increases

% Manganese (Mn) increases to achieve the strength increase

Chemical analysis

Carbon (C) 0.3% max.
Manganese (Mn) 0.1% max.
Phosphorous (P) 0.04% max.
Sulfur (S) 0.045% max.
Silicon (Si) 0.6% max.
Nickel (Ni) 0.5% max.
Chromium (Cr) 0.5% max.
Molybdenum (Mo) 0.2% max.
Copper (Cu) 0.3% max.
Vanadium (V) 0.03% max.

Mechanical properties

Yield 36 ksi (250 MPa) minimum
Tensile 70 ksi (485 MPa)–95 ksi (655 MPa)
Elongation (%E): 22% minimum
Reduction of area (%A): 35% minimum

FIG 5.7
A typical forged steel: SA-350 LM2

This is in common use for small forged steel fittings such as threaded pipe bends, caps and bushes.

Chemical analysis	Mechanical properties
Carbon (% C) 0.3% max.	Yield
Silicon (% Si) 0.15–0.3%	($R_{p0.2\%}$ proof) 36 ksi (250 MPa)
Manganese (Mn) 0.6–1.35%	Tensile 70–95 ksi (485–655 MPa)
Phosphorous (P) 0.035% max.	% reduction in area 22% min.
Sulfur (S) 0.04% max.	Impact strength at 46°C 20 J average
Copper (Cu) 0.4 max.	16 J minimum
Nickel (Ni) 0.4% max.	
Chromium (Cr) 0.3% max.	
Molybdenum (Mo) 0.12% max.	
Vanadium (V) 0.08% max.	
Niobium (Nb) 0.02% max.	

Hardness: Maximum 197HB

Note how:

- Proof strength ($R_{p\ 0.2\%}$) is used instead of yield
- %A is specified in preference to %E
- Impact strength (J) and hardness (H_B) are specified as they are important for a forged steel.

FIG 5.8
Carbon steels

- Carbon steel is almost all Iron (Fe) with carbon and only very small amounts of other trace elements (Si, Al, Mn, Mg) present

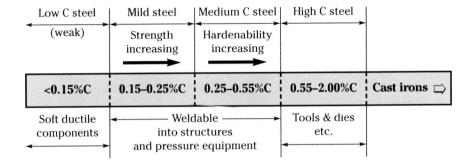

- Note how small amounts of carbon (C) produce significant effects in the steel's mechanical properties.

FIG 5.9
How mechanical properties are inter-related for carbon steel

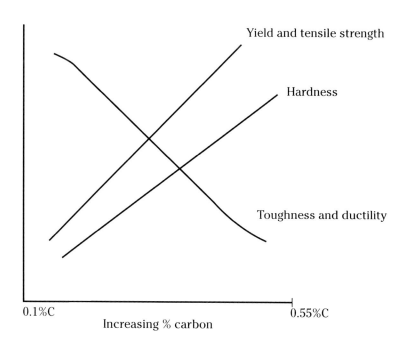

Remember the three main links between these properties for carbon steel

- Link 1: *Hard* material is *brittle* (low toughness)
- Link 2: High-strength (*strong*) material has *low ductility*
- Link 3: A *ductile* material is generally *tough* because its ductility *causes* toughness

5.4 Special materials

There are several areas of material usage where manufacturers like to develop their own materials, based on their experience. There is nothing inherently *wrong* with this, but be prepared to put in a little extra effort to understand the situation. Note the following guidelines when faced with manufacturers' 'own materials'.

- Try to identify the nearest published material standard to the manufacturer's own material. Use the elemental chemical analysis as a comparison first, then compare the heat treatment and any finishing processes.
- Make a careful comparison of the mechanical properties. Expect the 'own material' manufacturer to claim that their material is better than the nearest published standard. Ask how.
- For a material to be 'superior' to a published material, it is necessary that all the properties should be better. It is not always acceptable, for instance, to achieve better strength (yield or ultimate tensile strength (UTS)) at the expense of ductility or impact toughness or vice versa.
- Remember that there are other, less definable aspects to material properties than those common ones that you see in material specifications and standards. Fatigue and creep are specific examples; corrosion and erosion resistance are general ones. You can ask the material manufacturer to explain how they have taken these into account.
- The areas where you can expect to meet manufacturers' own material specifications in conventional fixed equipment are reasonably well defined. Practically, they are
  - large castings that are subject to high temperatures such as rotating equipment casings and steam valve chests
  - cast materials that have been developed specifically for increased corrosion resistance, such as valves for seawater and aggressive process service
  - forged materials that are designed to operate at temperatures above 650°C.

5.5 Material traceability

You will need to guard against the whole issue of material traceability looking, to you, like a paperwork exercise. This is dangerous, because fundamentally it is not. It is about *compliance*.

Traceability – to achieve what?

To achieve positive identification, so the buyers know they are getting the material that they have specified and that industry has spent tens of thousands of man-hours studying, understanding and improving. With this as the objective, the questions becomes one of how, in the world of real manufacturing industry with all its pressure and personalities, can such positive identification be achieved? The answer is in three parts, as follows.

- Material is fully examined, tested and classified at the source of manufacture – in the foundry, forge or mill.
- It carries *source* certification when it is sold.
- The system that follows this material to its final use is a *documentation* one. There *can be* some physical corroboration of what the documentation says.

Looking at this, you can see that it is essentially a documentation system which is given *solidity*, not exactly *validity*, by the existence of physical observations. Do not confuse this with an *administrative* system where documentation is more of an end in itself, because in material traceability, the documents only form the means to the end. This demonstrates the absolute necessity of a predetermined schedule of witness and review points for the SI to maintain the efficacy of the relationship between the documents and the pieces of material to which they apply. These witness points are shown in the ITP.

Figure 5.10 shows the *chain of traceability* which operates for engineering materials. Note that although all the activities shown are available for use (i.e. to be specified and then implemented), this does not represent a unique system of traceability suitable for all materials. In practice you will find several levels in use, depending both on the type of material and the nature of its final application.

Levels of traceability: EN 10 204

The most common document referenced in the material sections of ITPs is the European standard EN 10 204. It is widely accepted in most industries and you will see it specified in both American and European based contract specifications. It provides for two main 'levels' of certification: Class '3.1' and Class '3.2'. The highest level of confidence is provided by the '3.2', which requires that tests are witnessed by an independent third-party organisation. Class '3.1' certificates can all be issued and validated by the material manufacturer.

FIG 5.10
The chain of material traceability

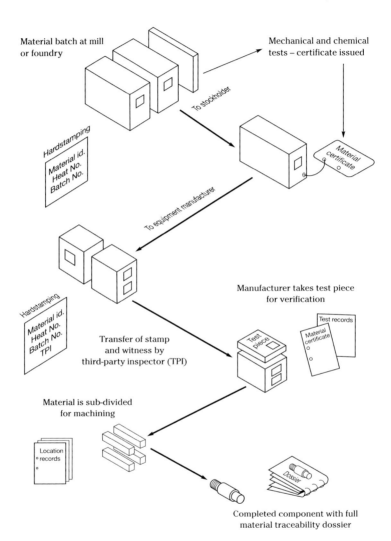

Retrospective testing

In cases where full traceability of material (as in Figure 5.10) is clearly specified, perhaps for code compliance reasons, there is little room for manoeuvre if you find at an advanced stage of manufacture that the chain of traceability is incomplete. This is not too uncommon. One solution is to do retrospective tests; here are some guidelines.

- The real objective is to identify and classify *fully* the material, not to carry out a few placebo tests to save making proper decisions.
- Some positive material identification (PMI) test methods cannot differentiate accurately between grades of some materials. They give an indication only. Other more expensive methods are more accurate.
- To be really sure about a material probably means cutting a test piece from the component in question. This is often actually much easier than the various parties (particularly the equipment manufacturer) think it will be.
- If you are not convinced that the original ITP has been followed correctly, then you are justified in requesting retrospective material tests. Check that you are supported by the text of the contract specification.

Responsibilities

The party responsible for correct material traceability should be defined in the contract specification. It is normally the main contractor but in practice, because of the structure of subcontracts, the responsibility may be delegated. As a SI you may be surprised by the way that material traceability requirements disappear or change within a chain of two or three subcontractors. There is no single explanation as to why this happens, but it often does. Here is a useful checklist of points related to the *responsibilities* of effective material traceability.

- First, check the text of the subcontract orders. This is normally where the traceability requirements have been 'mislaid'. There are few manufacturers that will do voluntarily what they are not contracted to.
- Check the role of the third-party (or authorised) inspection (TPI) body. The TPI is the nearest to the core witnessing activities at the foundry or mill, before material starts being cut, welded and machined. Help, and encourage, the TPI to pay extra attention to material verification at this stage.
- Monitor manufacture. The traceability of material during and after

cutting and machining is the task of the manufacturer. The TPI has less capacity for observation and influence once manufacture has started. Effective source inspection involves managing these parties in the best way at each stage of the manufacturing process.
- Investigate discontinuities in traceability. These are nearly always a 'system fault' rather than an isolated incident (ask any quality system assessor). You have to get to the *root cause* of discontinuities in the chain of material traceability, not just report that they exist.

5.6 Test procedures and techniques

Material testing techniques are well proven. They feature in many engineering reference books and are well documented in both American and European technical standards. Common destructive tests are used to determine the mechanical properties of a material and source inspection involves either witnessing the tests themselves (and so satisfy yourself of the accuracy and validity of the results), or more simply in just a *reviewing* capacity, that is, checking the test results on material certificates. The problem with material tests is that it is rarely the test procedures or activities themselves that reveal any potential problems with compliance. This is because most of the technical *risk* is not in the test, it is in the choice of specimens. It is perfectly feasible for a set of well-specified, executed and documented mechanical tests to be unrepresentative of the properties of the component from which the test piece was taken, if the specimens are badly chosen.

Do not expect to find many non-conforming results when watching the equipment and instruments during destructive tests in the laboratory; good manufacturers will have satisfied themselves of the test results before you arrive.

Test results – presentation

What should you expect to see on a material certificate? Normally this is quite straightforward; Figure 5.11 shows a typical example, in this case for an austenitic steel casting – a material broadly known as a $2\frac{1}{2}\%$ molybdenum steel. Although it may appear simple, a few points given below are worthy of note.

- **Acceptance levels**. The field of materials science is so wide that it does not make sense to try to anticipate specified values for either the chemical analysis or mechanical properties. Surprisingly, you will see a lot of material certificates that do not show the *specified* values;

instead they may just quote a material specification number (or sometimes not even that). Unless you are very experienced with a particular material, you should check with the material standard itself. Compare the figures carefully, making sure that the test results meet the specified acceptance values for the *particular* grade and heat treatment condition that you want. Be careful not to confuse minimum and maximum specified values.

- **Units**. In Europe, the majority of material certificates will use the international system of units (also known as SI units). You will see tensile strengths expressed in MPa, MN/m^2 or N/mm^2 – they all mean the same. Elongation (%E) and reduction of area (%A) are always expressed as a percentage figure. Note that tensile results are only relevant if the correct specimen size has been used. A good material test results document will show this; look for a symbol like '$L_o = 5.65 \sqrt{S_o}$' which indicates the specimen gauge length and diameter. Impact values are given in Joules (J) for SI units and foot-pounds (ft lbs) for US units.
- **Temperature classification**. Check the temperature at which the mechanical tests were carried out and compare it with that required by the standard. The material designation is normally a clue; for example, castings with special low-temperature properties have the suffix 'LT' after the material code. A series of digits may also be included to specify the actual temperature at which tests have to be carried out. If there is no explicit temperature information given, you are safe in assuming that all the tests and acceptance levels are referenced to ambient temperature (20°C).
- **Impact tests**. Not all materials have a specified impact value. Materials which are chosen for temperature applications above 150°C rarely need it, unless they are subject to shock loading conditions. For the alloy shown in Figure 5.11 the standard only requires impact tests if a low-temperature (LT) grade is specified.
- **Marginal results**. *Always* look carefully at marginal results, that is, chemical analysis or mechanical test results that are *just within* the acceptance limits, or fall exactly on the limit. Whenever you see a marginal result (look at the Mn content, and the proof strength (R_P) and tensile strength (R_m) results in the example), it is the time to dig a little more deeply. In practice you are almost obliged to accept the validity of a marginal chemical analysis result (unless you are willing to specify a re-test). For marginal mechanical test results, however, there are several steps that you can take, as follows.
    - Do a Brinell hardness test on the component (this can be carried

out in situ using a portable tester) and make an approximate conversion to tensile strength using the rule of thumb shown under the hardness testing part of this chapter. Be wary if you see results that indicate the material is significantly softer (lower strength) than the marginal test results suggest.
- If it is a graded material, look at the next grade down (lower strength) and compare the strength and the elongation values. Is there a big difference in the elongation values? This can give you a pointer. Note in the example how the marginal $R_{P1.0\%}$ and R_m are accompanied by an elongation of almost twice the minimum level. Although the material appears within specification, there is still a shadow of uncertainty. You can often see this pattern of results on steel sheet and strip material. This material is a good candidate for a re-test.

Tensile tests

This is the main destructive material test. European standard EN 10002-1 provides a detailed technical explanation of the technique. An equivalent American standard is ASTM A370.

From a source inspection viewpoint one of the main points of concern during tensile tests is the origin and *orientation* of the test specimens. Steels worked by all the main processes – forging, casting, rolling and their variations – have some directional properties, owing mainly to the orientation of the metal's grain structure. Large castings, which should in theory be homogeneous, still exhibit directional properties owing to the complex mechanisms of cooling and shrinkage. Tensile properties can vary significantly (perhaps up to $+25\%$) with the orientation of the test specimen with respect to the local grain direction. The shape of the load–extension characteristic can also vary. Many specifications, such as boiler and pressure vessel standards, specify clearly the orientation of tensile specimens, but some others do not. A few other guidelines may be useful to you.

- **Gauge lengths**. There are a number of different specimen sizes and gauge lengths. Make sure the correct one has been used for the material in question.
- **Defined yield points**. Make sure you know whether a material can be expected to show a defined yield point (note that there may be two, Re_L and Re_H but lower yield strength, Re_L is the important, commonly used one). It is almost impossible to tell this from the chemical analysis alone – look at the mechanical properties section of

FIG 5.11
Typical material certificate content and presentation

This is a '$2\frac{1}{2}$% molybdenum steel' austenitic casting material.

Chemical analysis

	C%	Si%	Mn%	P%	S%	Cr%	Mo%	Ni%
Specified	0.03 max.	1.5 max.	2.0 max.	0.040 max.	0.040 max.	17–21	2–3	10 min.
Actual	0.025	1.35	2.0	0.038	0.036	18	2.8	11

Mechanical properties

	$R_{p\ 1.0}$ (N/mm^2)	R_m (N/mm^2)	A (%)	Impact (J)	HB
Specified	215 min.	430 min.	26% min.	–	–
Actual (room temperature)	220	430	51%	–	–

the material standard to see whether it expresses a yield value, or resorts to a proof stress measurement.
- **Do not forget traceability**. Tensile specimens should be identified by documentary records and hardstamping. Piles of unidentified specimens in the test laboratory do not inspire confidence. If in doubt, ask for a re-test.

Tensile strength: easy rules of thumb?

Because of the large number of different grades in most material categories, reliable rules of thumb are difficult and there is a large degree of overlap between material types. Grades of cast iron, for instance can have tensile strengths ranging from 200 N/mm^2 to 800 N/mm^2. Carbon steels and stainless steels extend throughout a similar range. Unless you have a real tendency towards metallurgy (and most SIs do not), you should learn to rely on published standards. Express mild surprise when you meet SIs who have never even read them.

Impact tests

Impact tests are specified for materials that experience low temperature, shock loadings or both. There are several types of test, but the most common one is the Charpy V-notch test. This is described in the standards BS EN 10045-1 and ASTM E812. As with tensile tests the choice of specimen location and orientation is important. Figure 5.12 is equally valid for the location of impact test pieces for common material forms. Note that impact test results are inherently *less* reproducible than are tensile test results. This is due to the nature of the test itself – it relies upon a very accurately machined notch and is affected by inherent effects of material structure. Impact values also have great sensitivity to minor variations in grain size and precipitates, so small variations in heat treatment can cause different results. For these reasons, impact tests are always carried out in groups of three specimens and the results averaged. Note the following two key points.

- **Test designation**. Although you will meet the Charpy V-notch test more than 90% of the time, you may occasionally see Charpy U-notch or Izod tests specified. Note that there is *no conversion* between these impact results. If the wrong test has been used, the only option is to re-machine correct test pieces and repeat the test.
- **Units**. The Charpy impact value is given in Joules (J) or foot pounds (lbf). This is the energy absorbed by the specimen in breaking and it is specified as a *minimum* acceptable value. There may sometimes be a maximum quoted, mainly for alloys where ductility (as represented by reduction of area and elongation) can be falsely represented because of the propensity of the material to rapidly work harden during a tensile test.

Fracture appearance transition temperature (FATT)

The FATT test is an extension of the activities involved in the impact test and is simpler than it sounds. The objective is to determine the temperature at which the material will become brittle and break by a brittle fracture rather than a ductile fracture mechanism. This temperature is quoted as the *transition temperature*; it is essentially an indirect measure of the low-temperature impact properties of the material. The test is performed by doing a series of impact tests (usually Charpy) at $0°C$, $-20°C$ and $-40°C$ and recording the values.

A visual (and often microscopic) examination is then made of all the fracture faces with the objective of describing the fracture surface in terms of the relative percentage *areas* of ductile fracture and brittle fracture. As the test temperature becomes lower, the percentage of brittle fracture surface will increase. The FATT is defined as the test temperature at which these percentages *are equal*, that is, 50% ductile fracture/50% brittle fracture. You will see that there is room for some

FIG 5.12
Test specimen locations

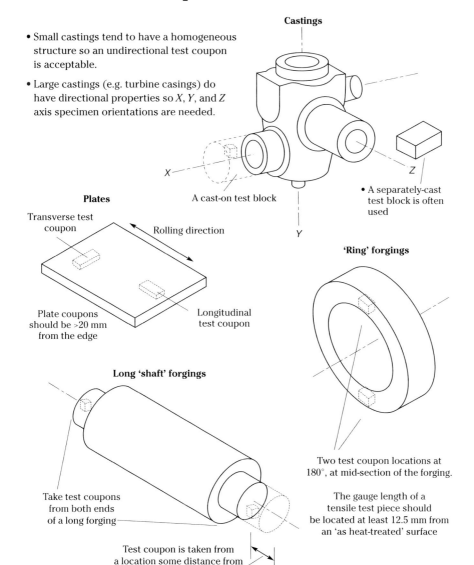

uncertainty in this test, so do not expect the FATT test to give results that are too precise.

As a guide, you should find that steel alloys which have high carbon, sulfur, silicon or phosphorous levels tend to have a relatively high transition temperature. Manganese and nickel additives have the opposite effect – they lower the transition temperature.

Hardness tests

Hardness tests are very quick and can be carried out in situ on a finished component as well as on a machined test piece. For in situ tests it may be necessary to lightly grind the metal to provide a good surface finish. The hardness test consists of pressing a steel ball, diamond or similar shape into the surface of the metal and then measuring either the force required, or the size of indentation for a particular force. Both are a measure of hardness of the material. There are several hardness scales, depending on the method used. The most common ones are Brinell, Vickers and Rockwell (B and C). Unlike impact tests, it is possible to convert readings from one scale to another. Figure 5.13 shows the approximate comparison scales for steel – note the abbreviations that designate the method used. More detailed data are given in the standard ISO 4964. Specific standards are available which describe the individual test methods – these are BS EN 10003 (for Brinell tests), BS EN 10109 (Rockwell) and BS 427 (Vickers).

The most frequently used hardness test for common forgings and castings is the Brinell test. The Vickers and Rockwell C tests are more suitable for harder materials. Note the following specific points on the Brinell test.

- **Terminology**. You should see a Brinell result expressed like this **226 HBS 10/3000**

    This is more easily read from right to left, as explained below.

    - 10/3000 shows the ball size used (10 mm). The '3000' is a 'load symbol', which is expressed as a factor (0.102) × the test force in Newtons. The force used depends on the expected hardness of the material.
    - HB denotes the Brinell scale. S (or W) shows that a steel (or tungsten) ball was used.
    - 226: This is the actual hardness reading. In laboratory tests it is determined by measuring the diameter of the indentation using a microscope and a special eyepiece incorporating a measuring

FIG 5.13
Comparison of hardness scales (for steel)

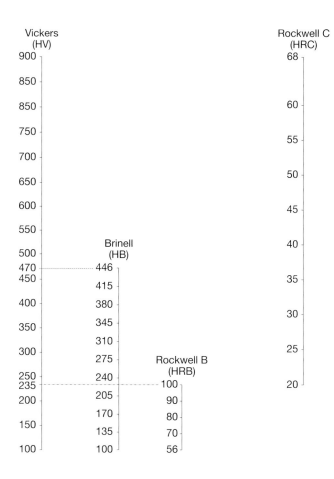

- Use this figure to make comparisons between hardness scales
- Treat the comparisons as approximate rather than exact
- It is often best to quote the HV hardness in your report – it is more commonly used and understood

- aperture. With portable hardness testers, the result is recorded as a function of the indentation force and displayed directly on a digital readout.
- **Accuracy**. For a laboratory test with a machined specimen, the test should be accurate to approximately $+2\%$ of the Brinell number. If you are using a portable tester it should be calibrated on a strip of test material before use. Even so, expect the accuracy to be a little less, perhaps $+3-4\%$. It is not advisable to rely on a single reading – many portable meters have an averaging facility that will calculate the mean of 10–20 readings.
- **Castings**. For in situ tests on cast components it is essential to grind off the surface layer where you plan to do the hardness test. If not, you will obtain misleading (harder) readings due to contamination and other surface effects.
- **A quick approximation** to tensile strength. For the Brinell test you can make an approximate conversion to the tensile strength using
  - UTS $(N/mm^2) \approx 3.39 \times$ Brinell hardness number (HB)

This can be useful as a quick 'order of magnitude' check when the tensile properties of a material are questionable. Be wary of the accuracy, however, because hidden factors such as work-hardening rates and grain structure can distort the conversion. It is not very accurate for cast iron. If in doubt during a source inspection, ask for a full tensile test.

Re-tests

Re-tests of material specimens are commonplace. Despite the efforts of technical standards to specify closely the methodologies of material testing to reduce uncertainties, the majority do make provision for re-tests if a material fails to meet its acceptance criteria during the first test. The philosophy of the acceptance criteria for re-tests is different for tensile and impact tests. Tensile tests have better reproducibility and most acceptance decisions are made on the basis of one or two results, whereas impact tests use an averaging approach. Figure 5.14 shows the way in which additional test specimens add cumulatively to the average. This approach is based more on statistics than metallurgy so, for re-tests, observe the following points.

- Check the material standard for the number of re-tests that are allowed.
- It is acceptable for components to be re-heat-treated to try to

FIG 5.14
Impact test retests

For impact tests (on pressure castings to BS 1504)

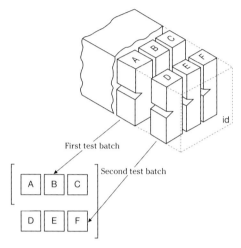

- If acceptance level = (say) 35J minimum. The material has failed the first test if the average (A+B+C/3) < 35J or if any one of A, B or C < 70% of (35J). A retest is allowed; take 3 more test pieces DEF from the sample: the material has still failed if the average {(A+C+D+E+F)/6} < 35J or if more than two are > 35J, or if more than one is < 70% of (35J).
- Then the component may be re-heat treated a maximum of twice and the above tests repeated. If it still fails, no further activities are allowed.

For tensile tests

If the first specimen fails the test

Two more specimens are allowed

If any <u>one</u> still fails

Two re-heat treatments allowed then re-test

improve poor mechanical properties. Pay special attention to the way in which the manufacturer plans to obtain the new test pieces (there may be no test bars left). It is advisable to positively identify test pieces with hardstamping before any re-heat-treatment. As the material has already failed a test, it is *unacceptable* to rely on unidentified test specimens. Make doubly sure that you ask the manufacturer about the origin of re-test specimens. What you do not want to hear is uncertainty. Note the agreed re-test piece location and orientation as well as traceability details on your NCR.

- *Always issue an NCR* if you find an incorrect material test procedure or an unacceptable test result. Let the re-tests or re-heat-treatment follow later.

Figures 5.15–5.17 provide some summary points you may find useful.

FIG 5.15
Key points: Materials of construction

Fitness-for-purpose (FFP)
1 The three main integrity/compliance criteria for materials of construction are
- mechanical properties
- temperature capability
- positive identification.

Material standards
2 Material standards often relate to a particular material *form* – forging, casting, plate or tube.
3 Acceptance guarantees are heavily dependent on technical standards. The situation is complicated by manufacturers who use 'own specification' materials – this book provides some simple guidelines for dealing with this.
4 Some material standards contain complicated tables and footnotes – these merit careful reading because they affect significantly the technical content.

Traceability
5 Verifying material *traceability* is an important part of the works inspection role. Try to see traceability as a control mechanism, rather than a documentation exercise. Expect, at times, to find this difficult.
6 There are various *levels* of traceability. EN 10 204 is the accepted standard covering the type of material certificates corresponding to the different levels.

Material tests
7 For tensile and impact tests the location and orientation of the test piece in the component is important. This is because material properties are *directional*.
8 You can convert between hardness scales (Brinell, Vickers and Rockwell) but not between impact strength scales (Charpy and Izod).
9 *Re-tests* are allowed if materials fail their first mechanical tests. Check for positive identification of the test specimens.

Inspection is about material *verification* – not material *selection*. Don't get too deeply involved.

FIG 5.16
Typical ITP traceability requirements

* It is essential to reference the applicable standard and acceptance levels ↓

* Note the role of TPI for traceable 'statutory' material ↓

high ↑ level of material traceability ↓ low

Step No.	Operation	Inspection Points M / C / TPI	Certificate Requirements	Comments
	Boiler Header Tube			
1.	Transfer of marks	W R W	EN 10 204 (3.2) certificates	
2.	Mechnical tests	W W W		
3.	Chemical analysis	W W W		
4.	Hydrotest	W W W		Note any
5.	Tube NDT	W W W		NCRs in
6.	Documentation review	W R R		this column
	Safety Valve Spindle			
1.	Mechanical tests	W W R	EN 10 204 (3.1) certificates	
2.	Chemical analysis	W W R		
3.	NDT	W W W		
4.	Documentation review	W W W		
	Boiler Structural Steel Work			
1.	Mechanical tests (sample)	W W R	EN 10 204 (3.1) certificate for carbon steel	
2.	Chemical analysis (sample)	W R R		
3.	Document review	W W R		
	Steam Pipe Expansion Joints			
1.	Document review	W R –	Certificate of conformity	

↑
*Non-traceable 'catalogue' item

↑
*Spell out clearly which level of certification is required

W = Witness point M = Manufacturer C = Contractor TPI = Third-party inspector
R = Review
↑* Include a key so there is no misinterpretation.

FIG 5.17
Job well done: Material tests

1. *For all tests*. Make sure you know the location, orientation and number of test pieces that were taken.

2. Tensile tests. Standard symbols should appear on material test certificates. These are:
    Re: Yield strength (Re_L is the 'lower' yield strength, Re_H is the 'higher' yield strength)
    $R_{P0.2}$: 0.2% proof strength
    R_m: UTS or 'tensile strength'
    $A\%$: Percentage elongation of gauge length

Check that:
- The correct test temperature has been used. Some materials are tested at elevated temperatures.
- The correct Re or $R_{P0.2}$ measurement has been taken (look at the relevant standard).

3. Impact tests
- Check the specimen size and notch configuration. There are several different types and it is not possible to convert the results.
- Try to examine the broken specimens so you can describe the fracture surfaces in your report.
- Check the contract specification to see whether it specifically asks for a FATT test.

4. Hardness measurements
- Make sure the correct scale (HV, HB, HRB or HRC) is used.

Chapter 6

Inspecting welding and fabrication

6.1 Welding inspection

The objective of source inspection of welds is *code compliance*. In turn this should ensure the *integrity* of the weld; an explicit objective of the relevant statutory codes and standards for all types of engineering equipment incorporating the jointing of metals. It is also an implicit requirement for non-statutory equipment because of its impact on the safety aspect – there is a direct link to a manufacturer's duty of care to provide inherently safe equipment, and to their formal product liability responsibilities. Your job as a source inspector (SI) is to *verify* this code compliance, using it as a focus in your inspection activities, reports and NCRs.

The scope of SI weld inspection activity divides roughly into three areas, as described in the following three sections.

Check for the correct welding technique

Every welding application has a welding technique that is the most suitable, so a well-established methodology has developed for documenting and controlling welding techniques. This comprises the use of weld procedure specifications (**WPS**), weld procedure qualification records (**PQRs**), and the testing and approval of welders themselves to perform the WPSs. The purpose of this methodology is to ensure the correct type of welding technique is specified, and then to control its implementation.

Check for the correct tests

There is only a limited amount that you can conclude from visual examination of a weld. Destructive tests and non-destructive evaluation (NDE) are required to look properly for defects that affect integrity.

Non-destructive techniques are essentially *predictive* tests. Because each technique can only detect certain types and orientations of defects, the choice of technique is critical – it always depends on the application. As an SI you need to pay attention to the suitability of the NDE technique being used in order to understand the accuracy and validity of the results that will be achieved.

Acceptable levels of defects

Note the reference to *acceptable* levels of defects. It is part of the SI role to compare defects found with the correct defect acceptance criteria for the equipment in question. Acceptance criteria are documented in many engineering codes and standards but not always in the component-specific way that would be most helpful. Be prepared to exercise some judgement on the applicability of defect acceptance standards to the component you are inspecting.

Welding techniques

No welding activity is perfect. Forget the notion that a closely controlled welding procedure backed up by a series of 'proving' tests will produce a good weld every time. Most welding of ferrous materials for static equipment fits into one of four main categories. These are, using the ASME definitions

- shielded metal arc welding (SMAW)
- gas metal arc welding (GMAW or MIG)
- gas tungsten arc welding (GTAW or TIG)
- submerged arc welding (SAW).

There are others classified as *automatic techniques*, which include plasma and electron-beam welding and similar. Together, these manual and automatic techniques will account for perhaps 95% of the welding methods that you will meet during source inspections of static equipment. Each technique has its own set of uncertainties that can (and do) result in defects. It is difficult to place the techniques in order of the *risk* of producing defects and there is little firm evidence to conclude that manual techniques produce more defects because of the existence of a human operator. Sometimes the opposite is the case. Butt welding techniques for small-diameter tubes, for instance, are often really only semi-automatic techniques. Critical weld variables may be automatically controlled but the process still requires an operator and defects are still produced. You may find that *automatic* does not necessarily mean *better*.

Why do welding defects occur?

The character of weld defects provides some information as to their origin. Figure 6.1 shows an overview of how weld defects occur, or more specifically *what* causes them. On balance, this figure holds good for any of the main ferrous welding technique groups, whether automatic or manual, and has key implications for the way that you should organise your priorities during source inspection of welding. Note the following guidelines.

- The *main* cause of weld defects is poor process conditions during the welding activity.
- Operator error is directly responsible for nearly one-third of defects, even when the process conditions are all correct.
- Errors caused by incorrect material and electrodes, or an unsuitable technique, are not too common. Those that do occur are usually due to poor manufacturing practices rather than technical uncertainties.
- Although documentation is involved at various stages of the control of a welding process, the chances of a pure documentation fault being the root cause of a defect are actually quite small (note the shaded areas in the figure). Most of these consist of straightforward errors of communication where the wrong work instruction has been passed to the shop floor and almost all will have occurred *before* an SI became involved.
- Figure 6.1 can only be an approximation because it applies to all four main technique groups, which are quite wide. Expect the proportion of technical (i.e. non-operator) problems to be slightly higher for those techniques that have a greater level of inherent technical risk. These are
  - materials with heavy wall sections (>75 mm)
  - a parent material of ferritic stainless steel ($>16\%$ Cr) or unstabilised (no Ti or Nb) austenitic stainless steel
  - any structural butt weld involving dissimilar materials: particularly carbon steel-to-stainless steel
  - any welding activity where the parent materials (or the electrodes) are of uncertain origin or do not have positive identification.

There is a fair degree of reproducibility in weld defects, so you *will* see the same problems repeating themselves. It should not take long for

FIG 6.1
How weld defects occur

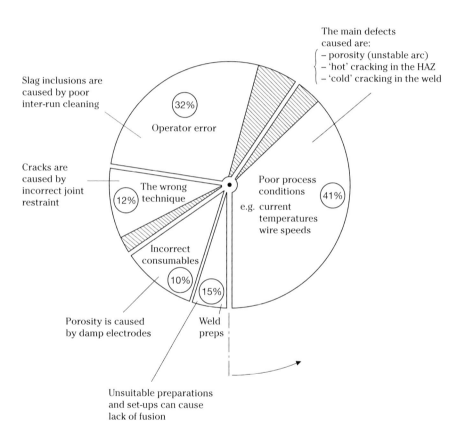

them to fall into an approximate pattern, broadly in line with that shown in Figure 6.1.

6.2 Types of welds

There are a small number of fundamentally different types of welds with a few variations of each type. Weld type is of key interest to an SI mainly because of the way it affects the NDE requirements (or lack of them) that attach to particular types. Some of these linkages are specified in engineering codes EN, ASME and so on, whereas for other types of equipment greater reliance is placed on what is generally accepted as good engineering practice. The most important weld type, which has the most stringent non-destructive testing (NDT) requirements, is the *full penetration weld*. This is in common use in pressure components and other static equipment alike.

The full penetration weld

The most common definition of a full penetration weld is a weld that extends completely through the thickness of the components joined. Its purpose is to transmit the full load-carrying capacity of the components without any weaknesses.

Full penetration welds are normally the most important ones in an engineering component or structure. This is a further reason why they should be subject to the most stringent levels of NDE; in order to keep risk to a minimum. You should always be on the lookout for full penetration welds. Figure 6.2 shows some common examples, and their application.

Other weld types

Other types of welds which do not meet the full penetration criteria normally have less stringent NDE requirements. Fillet, lap and seal welds are the most common but there are many other types, some of which are often mistaken for full penetration welds. Figure 6.3 shows some examples that you will meet.

Weld heat treatment

Heat treatment is an important activity in obtaining desirable properties in an engineering material. You can think of welds in much the same way; because of the high local temperatures imposed during the welding

FIG 6.2
Full penetration welds – common applications

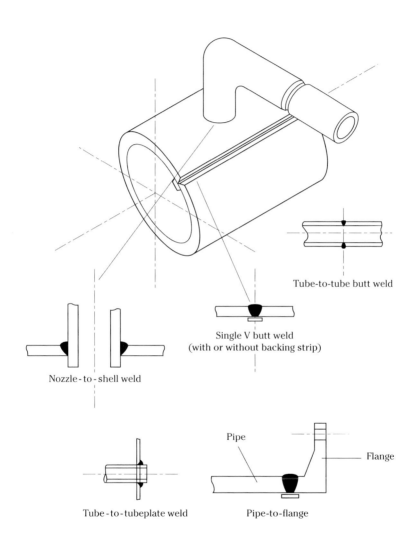

FIG 6.3
Other weld types

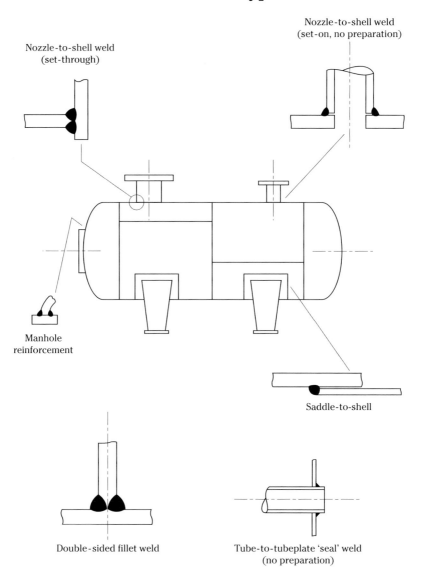

process, some type of heat treatment is often required. There are basically two sorts.

- **Preheat.** Pre-weld heating is performed on the prepared edges immediately before welding.
- **Post-weld heat treatment (PWHT).** This can be either the local type, which just reduces the cooling rate after welding, or a fully controlled stress relief carried out in a furnace after welding.

The purpose of these activities is to prevent cracking in the weld material and the heat-affected zone (HAZ), which is an important consideration whether manual or automated techniques are used. It is a consideration not just at the initial welding stage but also during repair welding, when a defect is found at a later stage of manufacture. Expect to see pre- and post-weld heat treatments shown on the ITP for fabricated items. If it is not, you should check to confirm that it is indeed not required. You can do this by checking in the applicable material standard or construction code.

Verification that the correct weld heat treatment has been carried out is part of the source inspection role. In terms of *priority* it is best to concentrate on whatever PWHT requirements are specified. Bigger and better weld defects are caused by incorrect PWHT rather than by poor pre-heating. We will see later in this chapter the importance of weld test pieces in verifying the final properties of the weld and HAZ.

6.3 Welding documentation

Welding is associated with a well-defined set of documentation. The term *controlling* documentation is a slight misnomer – the documentation cannot *control*, as such. It is better to think of this documentation as comprising a part (albeit an essential part) of the mechanism of controlling and recording. The documentation set also fits neatly in with the relevant requirements for statutory equipment and provides a common-sense way of working, which has become generally accepted for all types of fabricated equipment.

The objectives of this documentation set are to

- **specify** a particular weld method to be used
- **confirm** that this weld method has been tested and shown to produce the desired weld properties
- **ensure** that the welder who performs the welding has proven ability.

Terminology can be confusing. Technical standards such as EN and

FIG 6.4
Welding: controlling documentation

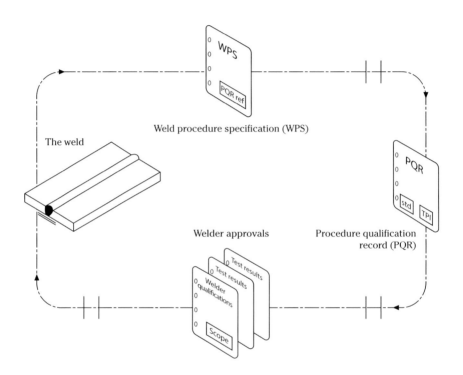

ASME refer to the need for the review of weld *procedures* and of the need for weld *testing*. These are fine as general definitions (they are well understood) but it is better if you refer specifically to those individual documents that make up the set. This reduces the chances of misinterpretation. Figure 6.4 shows the documents, and how they are related to the weld in question, and each other. The example is for a simple single-sided butt weld of a type commonly found in pressure vessels. Note that it has a backing strip – this is an acceptable configuration where one side of the weld is inaccessible. This is defined as a full penetration weld and could be added therefore to the examples shown in Figure 6.2.

The weld procedure specification (WPS)

The WPS describes the weld technique. It is a technical summary of the relevant weld parameters and detailed enough to act as a work instruction for the welder. The WPS is prepared by the manufacturer and the general principle is that each weld type should have its own WPS, containing details of

- parent material
- filler material/electrode type
- the weld preparation
- fit-up arrangement
- welding current, number of passes, orientation and other essential variables
- back-grinding of the root run
- pre- and post-weld heat treatment
- the relevant procedure qualification record (PQR).

The procedure qualification record (PQR)

Different standards use different terminology for this, but the term PQR is commonly used as a generic meaning. The PQR is the record of welding and testing of a specific weld, analogous to the 'type-test' carried out on some items of equipment. A test weld is performed to the preliminary WPS and then subjected to an extensive series of non-destructive and destructive tests to determine its physical properties. These include

- visual and surface crack detection
- ultrasonic or radiographic examination
- destructive tests (normally by bending to test the strength characteristics of the welded joint)
- hardness tests across the weldment and HAZ regions (done on a polished and etched macro-sample). Hardness gives an indication of heat treatment and metallurgical structure. The range of hardness readings across a particular sample can identify factors that lead to an increase in the risk of cracks developing in the future.

A weld test leading to the issue of a formal PQR is generally witnessed by an independent third-party organisation. If you look at a typical set of welding documentation and the relevant technical standards, you will see that not every weld (or every WPS) *necessarily* needs its own dedicated PQR. Note also that PQRs are unlikely to be job-specific (but

WPSs are). Most common weld types have been tested over the years and, as long as the essential weld variables do not change, PQRs do not go out of date.

Welder qualifications

Conventional wisdom is that welding ability is *non-transferable*. This means that the ability of a welder to weld correctly, for instance, low-carbon steel does not (rather than *may* not) mean they can do the same on stainless steel. Weld technique, position and material type all have a significant effect on ability. This is a harsh concept, but one which you are obliged to follow. As with the WPS situation, the requirement for qualified or *coded* welders has spread from its origin in the manufacture of pressure vessels and other statutory equipment. You will therefore find it a generally requested concept in most pressure and structural static equipment contracts.

A welder is tested to a specific WPS (or a range of them depending on the scope) and the weld is then tested in a similar or identical way to that used for the PQR. Approval is confirmed by the welder being awarded a personal certificate, which includes a photograph. The test results and certificates are normally certified by a third-party organisation.

The problems of mismatch

It would be nice if the technical scope of every particular WPS was matched exactly by a corresponding PQR, with the welder qualified to precisely the same weld arrangement (and the tests) used in the PQR. In practice, it does not always happen quite like this. Both EN and ASME standards allow a certain leeway in the essential variables of a weld, both between the WPS and the PQR, and between the PQR and a corresponding welder qualification. It is a little misleading to think of these 'gaps' as a mismatch – there is certainly no *conflict* between the technical requirements of a WPS and its nearest corresponding PQR. What it does do, however, is introduce a certain level of risk. This is one reason why the common set of technical documentation accompanying welding activity cannot provide *full* control of what is happening.

These gaps between WPS, PQR and welder approval requirements can cause you problems as an SI. You will not find it easy to recognise instantly whether a WPS and PQR correspond or not. Unfortunately the 'ranges of approval' are quite large – the standards contain complex matrices explaining which parent material groupings and material thickness differences are allowed. It is therefore necessary to *check* the matching of the WPS, PQR and weld approvals during a source inspection. This is an essential part of the mechanism for ensuring code compliance.

The technical coverage of welding and NDT standards is extensive. There is reasonable uniformity of approach between European and American (ASTM) sets of standards but expect also to find some duplication. The range of ASTM standards, many of which are referenced directly by the ASME pressure vessel codes, are very practical 'doing' standards. The standards in the European range seem to adopt a more technological approach. In large contracts you might find a mixture of both systems.

The role of weld acceptance criteria

Weld acceptance criteria are normally given in the *application* (construction) code for the equipment being manufactured. You cannot expect, however, that the definitive and prescriptive defect acceptance criteria included in codes for pressure vessels and statutory items will be available for other equipment; sometimes the inspector's judgement is required. Try to follow these general guidelines on acceptance criteria.

- Look carefully at their content – they contain good accumulated knowledge.
- For *coded* pressure vessels and statutory equipment you should apply rigorously the acceptance criteria laid down in the construction code as long as the *applicability* of the standard is clear.
- For other static equipment, put some effort into checking the applicability of the criteria to your particular piece of equipment. Do not fight the *content* of the acceptance standards.

but

- There will always be room for judgement and interpretation, so be prepared for this.

Welding ITPs

It is rare to find an ITP without some welding and NDT content. There are normally only a few lines for each type of welding activity, but this may be repeated many times through the ITP. The best example of this is for a pressure vessel, in which individual seam and nozzle welds have their own lines of the ITP.

Figure 6.5 shows typical ITP activities for a source inspection. For

FIG 6.5
A typical 'welding part' of an ITP

Step no.	Operation	Reference documents	Inspection points			Certification requirements	Record no.
			M	C	SI		
1	Weld Procedures	WPS/PQR	R	R	R	ASME VIII/IX	XX/Y
2	Welder approvals	ASME IX	R	R	R	ASME IX	XX/Y
3	10% RT	ASME VIII	R	R	R	Record Sheet	XX/Y
4	100% MT	ASME VIII	R	R	R	Record Sheet	XX/Y
5	Visual inspection	ASMR VIII	R	W	W	Record Sheet	XX/Y
6	Document review	-	R	R	R	-	XX/Y

W = Witness point
R = Review
M = Manufacturer
C = Contractor
SI = Third party (or clients) source inspection organisation

best effect you should do these while in the works and not by reviewing the documents at your office later. Note that there are few witness (W) points shown in the ITP. This is often the case for non-statutory fabricated equipment. It is not an ideal state of affairs; some witnessing is probably necessary, if only to stop a documentation culture taking over. When you do source inspection in a manufacturing works, make a point of trying to witness the key welding and NDT activities. Aim for the following

- weld preparations, particularly for thick steel sections (say > 20 mm); do a visual inspection *then* compare the set-up with the WPS
- circumferential (mis)alignment on head-to-shell joints before welding for all types of vessels; BS/EN, ASME and all other recognised vessel codes quote maximum limits
- grind-back of double-sided butt welds
- magnetic testing (MT) examinations in progress (because it is easier to get it wrong than with dye penetrant techniques)
- repairs of any type.

Above all, when you are in the works, *inspect the hardware* not just the paperwork. The fact that you are reading this book shows that you are an inspector and not an administrator.

6.4 Test procedures and techniques

Witnessing tests is not the same as having to carry them out. As an SI your objective is to possess just the right level of knowledge to enable you to verify specification/code compliance without wasting time or money (either your own or the manufacturer's). As a result, you cannot expect to achieve the level of feel for welding and NDE techniques that an experienced technician or operator will possess. Your job is to concentrate on *verification*, not to act as unpaid technical advisor to those performing the tests.

Checking weld preparations

Poor weld preparations can cause poor welds. A source inspection programme for equipment containing load-bearing welds should make provision for inspection of various stages of welding in progress, so make a point of being in the works to inspect some of the major welds at the preparation stage. The same principles apply for butt and nozzle welds.

Start with weld joint *design*. Check that the preparation design complies with the fabrication drawing, the WPS/PQR and any applicable code (EN/ASME), and so on, in that order. Pay particular attention to the weld preparation angles, and to the root gap or backing strip, as applicable. Pressure vessel codes show typical weld preparations that are acceptable for coded vessels, but do not expect these standards to be fully prescriptive as to whether *your* particular joint design is allowed. Sometimes you will have to use a bit of interpretation. The situation is normally clearer for butt welds than for nozzle welds. If in doubt you can issue a NCR, asking the manufacturer to demonstrate compliance with the relevant standard, then see how they respond.

The next step is to inspect the *prepared plate edges*. It is important that defects are removed before welding begins. Any defects will remain in the weldment, often at the edge of the HAZ and be a prime source of initiation for cracks. These are normally planar defects, so they may not be easily found by radiographic testing (RT). Make sure of the following points.

- Flame-cut plate edges and sheared edges should have been ground back (before the preparation angles are machined) by at least one-quarter of the plate thickness ($t/4$), preferably more, to remove burnt and work-hardened areas.
- MT or penetrant testing (PT) checks are performed on the prepared

FIG 6.6
How to check weld preparations

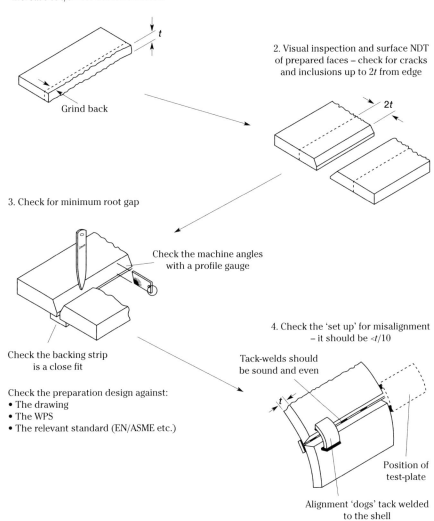

edges to look for cracks or inclusions. Do this on both sides of the plate, up to a distance of at least $2t$ back from the prepared edge (see Figure 6.6). Do not accept any crack-like 'indications' as acceptable at this stage – the only solution is to grind out and re-machine the preparation. Then repeat the check.
- For nozzle welds, be wary of poor surface finish on the outside of the nozzle. If it is a forging, remove all millscale to obtain a good clean surface before welding.

Finally check the weld set-ups. Particularly on large fabrications, physical alignment of the joint set-up can be a difficult task. The proper time to do the check is *after* all the necessary tack welds, braces and joint 'dogs' have been fitted; that is, the joint is fully ready for the first weld seam. Check the soundness of the tack welds and that all slag has been cut out. Look for a uniform root gap all along or around the weld. For nozzle welds, check that adequate bracing has been fitted to prevent the nozzle 'pulling over' during welding. You will have to use experience here, or ask the manufacturer. It will not be on the WPS, but it might be on the fabrication drawing. Do not lose sight of your objectives – you are looking for a weld set-up that will give a sound full penetration weld. It is not always essential to witness the welding activity itself, you can concentrate instead on a thorough examination between weld runs, and then again when the seam is finished.

Visual inspection of welds

It is safe to say that all welds should be subject to visual inspection. Strictly, it is a part of the NDE process – it plays a part when checking the location of radiographs and during the ultrasonic tests on the parent plate or completed welds. The way in which you approach visual inspections is more important than first appears. This is largely because the visible appearance of welds is easily commented upon by users, site engineers or inspectors after the equipment has arrived at its construction site. You can expect all manner of opinion, some of it informed and some not, on how the welding looks. For this reason, it is best to adopt a thorough approach to the way that you *report* your visual inspections of welds, and any defects that you find.

The best way is to use a *checklist*. This will provide a clear and simple way to present your findings in your report. It will help others (and occasionally yourself) to differentiate visible observations that truly affect code compliance from those that are merely cosmetic.

FIG 6.7
Job well done: weld inspections

1. The big objective

The main objective of weld inspection is to prevent defects that will results in cracking, either immediately after welding or later. As a source inspector you play a major role in this.

2. Metal-v-paperwork

Having the correct paperwork is not the same as having a good weld. Incorrect or missing paperwork may be a non-compliance but it doesn't mean that the weld is not (or is) correct.

3. Indications and defects

Remember that indications (and flaws, imperfections and discontinuities) are only classed as defects if they are outside the criteria of the application code to which the component is being constructed (ASME B31.3/VIII-I etc.).

4. What about contractors' weld inspectors

Most welding contractors employ their own weld inspectors whose job it is to hold various certificates indicating that they know how to inspect welds (read that carefully). Many are ex-welders and so have a natural tendency to inspect based on a reflection of the standards that they used to weld to when they were welders. As a source inspector, responsible for code and specification compliance, but one step removed from the manufacturing process, this provides you with an interesting dilemma. All you have to do is decide whether these weld inspectors are part of the solution, or part of your problem.

Chapter 7

Inspecting non-destructive examination (NDE)

Almost every source inspection involves welds in ferrous materials. Welding (and its metallurgy) and its subsequent non-destructive examination (NDE) are well-developed engineering disciplines with the result that levels of staff competence and qualifications are high. Although manufacturers' welding inspection technicians and their opinions abound during source inspections, you sometimes find that they prefer to wait for the source inspector (SI) to make the decision.

The objective of source inspection of welds is *code compliance*. In turn this should ensure the *integrity* of the weld, an explicit objective of the relevant statutory codes and standards for all types of engineering equipment incorporating the jointing of metals. It is also an implicit requirement for non-statutory equipment because of its impact on the safety aspect – there is a direct link to a manufacturer's duty of care to provide inherently safe equipment, and to his formal product liability responsibilities. Your job as an SI is to *verify* this code compliance, using it as a focus in your inspection activities, reports and NCRs.

7.1 Surface crack detection

A disproportionate amount of cracks and defects are to be found on the surface of cast, forged and cold-worked materials compared to those that are concealed in the body of the material (with the exception of the weld root, which *is* a common source of defects). This is due mainly to the way that the component material is formed, worked or/and welded. There is little more that can be done technologically, than is done already, to stop such defects occurring. Although much study and work has been done to develop specialised, often esoteric surface crack detection techniques, the majority of manual methods used in manufacturing works still use simple dye penetrant testing (PT) or magnetic particle testing (MT) techniques.

In a practical source inspection situation these techniques provide *only* an enhanced visual assessment of the surface. Frankly, they will show you very little that a close visual examination under bright illumination will not. They are useful, however, for identifying surface indications and generally making the job of finding defects that much easier. They are also quick and relatively simple to perform. As with all NDE techniques, their effectiveness depends on factors such as the surface finish of the material, orientation of the defects, and a certain amount of skill and familiarity.

Limitations

As an SI, you have to treat surface NDE results with some caution. PT and MT techniques will detect cracks of visible size which, if left unattended, can cause failure of the material – but there is a limit of *very approximately* 0.1 mm (100 µm) minimum defect size that will be detected when using these methods in a practical source inspection situation. This 100 µm is *larger* than the 'critical crack size' for many materials (critical crack size is that which is considered large enough to propagate under normal working stresses and hence cause failure). Treat surface NDE seriously, certainly, but try to complement the results with an understanding of the metallurgical realities of the mechanisms by which materials actually fail. This means take care.

Dye penetrant testing (PT) inspection

This is the simplest and best-known test for surface cracks. There are several standards explaining the technique and showing typical defects (see Figure 7.1). PT has the advantage that a visual record is available so indications can be photographed. Despite it being such a simple technique, you will still see it done incorrectly. Figure 7.2 shows the correct method. Three aerosols are used – the cleaner is a clear liquid, the penetrant is red and the developer is white. Note the following PT guidelines.

- PT is equally effective on all metallic materials, but is sensitive to surface finish so sharp edges and weld spatter will give false indications. As a guide, a normal 'as-cast' finish should be acceptable after fettling all rough edges. Anything rougher may need light grinding.

FIG 7.1
Surface crack detection

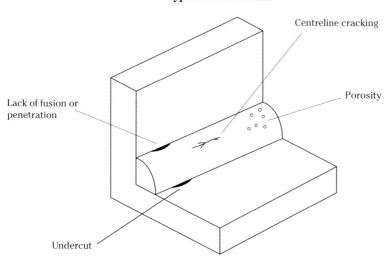

Typical weld defects

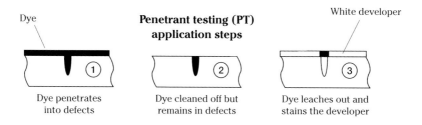

Penetrant testing (PT) application steps

- Make sure the correct cleaning steps are carried out, as shown in Figure 7.2. It is not advisable to scrub or rinse the surface to remove the red penetrant before applying the developer. This may wash it out of surface defects and give poor results.
- The most common error is not allowing the red penetrant enough time to penetrate. At least 15 min is required.
- Surface cracks will show as hard red lines on the surface of the white developer. Surface porosity will show as a group or chain of small red dots.
- Edge laminations on weld-prepared or sheared steel plate will show

FIG 7.2
Dye penetrant testing – the correct method

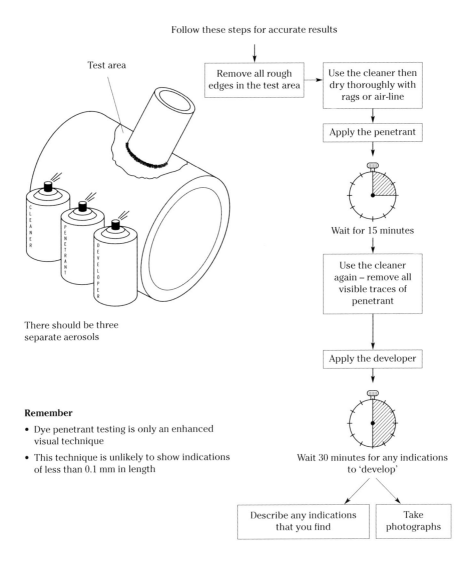

Remember
- Dye penetrant testing is only an enhanced visual technique
- This technique is unlikely to show indications of less than 0.1 mm in length

up as thin, often indistinct, crack-like indications orientated along the cut edge. On sheared plate, you may also see transverse cracks caused by excessive work-hardening of the plate. It is not possible to do a successful PT test on a flame-cut edge.
- If you find defects, take photographs to include in your inspection report.

PT techniques are regularly used on complex components such as heat exchanger tubesheets and pressure equipment headers where the number and arrangement of tube stubs and nozzles makes MT difficult. It is difficult to identify any areas where PT could not be used as an alternative to MT if required.

Magnetic particle testing (MT)

Magnetic particle testing works owing to the magnetic properties of the material being examined. A magnetic flux is passed through the material and the surface is sprayed with a magnetic medium or ink. An air gap in a defect forms a discontinuity in the magnetic field, which then attracts the magnetic filler in the ink and makes the defect visible. Large areas of material or weld can be tested without too much preparation, so it is commonly used on large vessels and fabricated structures. Theoretically, because the magnetic field penetrates some distance below the surface of the material, this technique can indicate the presence of sub-surface defects, as well as true surface defects. In practice this is difficult – an experienced MT operator is required to spot anything less distinct than a pure surface defect. It is best to take the view that MT, like PT inspection, can only provide an enhanced visual assessment.

There are several types of magnetic medium that can be used: dry red or blue powder, black magnetic ink and fluorescent ink viewed under ultraviolet light are all in common use and do basically the same job. There are also several methods of applying the magnetic field to the component under test. Figure 7.3 shows the common arrangements. Note the following guidelines.

- MT cannot be used on materials that are non-ferromagnetic. These include most austenitic stainless steel and non-ferrous materials and alloys.
- MT is better at detecting non-metallic surface inclusions than PT. For other defects its resolution is about the same.
- It is difficult to take good photographs of defects found by MT. Make sure you identify any defect during the test itself and note its

FIG 7.3
Magnetic particle crack detection – guidelines

Methods
1. Black magnetic ink (shows up better with white contrast paint).
2. Flourescent (under U.V.) ink – good, but needs a dark enclosure.
3. Dry powder (red or blue) – difficult to see small defects unless the surface is ground.

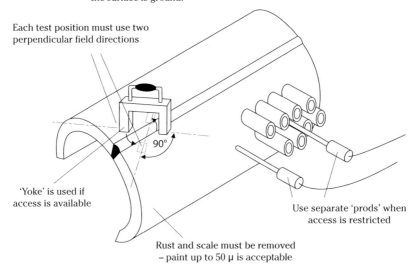

Each test position must use two perpendicular field directions

'Yoke' is used if access is available

Use separate 'prods' when access is restricted

Rust and scale must be removed – paint up to 50 μ is acceptable

Check the magnetic field using a tester

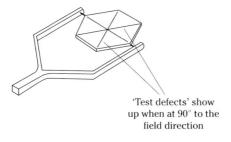

'Test defects' show up when at 90° to the field direction

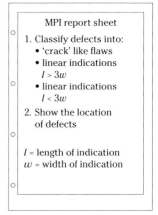

MPI report sheet
1. Classify defects into:
   - 'crack' like flaws
   - linear indications $l > 3w$
   - linear indications $l < 3w$
2. Show the location of defects

l = length of indication
w = width of indication

FIG 7.4
Defect types

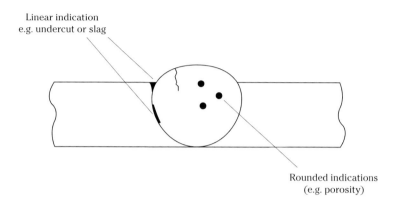

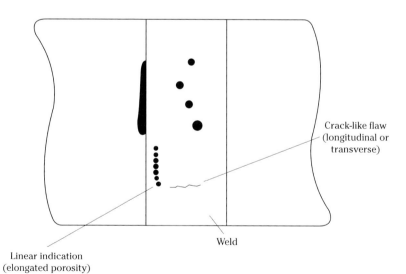

position on a sketch for inclusion in your inspection report. *Describing* the defect is important. There are three specific categories of defects that can be detected by MT: *cracks* (you will often see these described more generally as *crack-like flaws*), *rounded indications* and *linear indications* (see Figure 7.4). Your source inspection report should use these specific terms.
- There are two ways of applying the magnetic field – either permanently (the current is kept on while the ink is applied) or temporarily (the field is applied, removed and *then* the ink is applied); both are acceptable. In the second method the field is maintained by residual magnetism for sufficient time to apply the ink and identify any defects. If in doubt look at E709 or E1444.
- The technique is most sensitive when a flaw is orientated at 90° to the magnetic field direction. When the angle falls below 45° the sensitivity becomes poor.
- The most common technique error is not using *two* perpendicular field directions. This totally negates the usefulness of the test, as two directions are essential in order to locate defects in various orientations. Use the field tester, which contains test defects to ensure that the field strength and ink properties are good enough.
- Beware of 'phantom indications' in corners and around sharp changes of section, particularly in castings. The correct terminology for these is *non-relevant indications*.
- A paint coating of <50 µm will not affect significantly the sensitivity of the technique, so a thin coating of white contrast paint is often used to make the indications more visible

7.2 Volumetric NDE

Volumetric NDE is the generic name given to those non-destructive techniques that identify defects within the body of a material or component rather than on the surface. You cannot properly find and identify surface defects using volumetric techniques. They are more complicated than surface techniques and have greater variety. As an SI your objective is *familiarity* with the two main techniques, radiographic (RT) and ultrasonic testing (UT), and how they are applied during source inspections, in particular on structural welding of pressure vessels and other statutory equipment.

Which technique – RT or UT?

Ultimately, this is a question for you (and maybe your client). RT and UT are very different methods and there is little commonality in their method of use, conditions under which they work best, or their capability in finding specific types of defects. Broadly, however, they both do the same job.

Much of what can be considered conventional wisdom relating to volumetric NDE comes from pressure vessel construction codes. ASME codes allow both, but traditionally prefer RT, because of the permanent records of the results. Manufacturers are, however, increasingly using UT as an alternative on structural welds, with the cautious approval of purchasers and code bodies. As an SI you should follow any agreement made between the manufacturer and your client (the purchaser). This may only manifest itself across several levels of documentation, namely, technical specification clauses, construction code preferences and broad contractual references to 'correct standards' or suchlike.

Ultrasonic testing (UT)

There are four main areas where UT techniques are used: steel plate, castings, forgings and welds.

UT of plate material

Ultrasonic testing is used to check plate for *laminations*, extended flat discontinuities between layers of the rolled material. Plate for general structural purposes is not normally subject to a full lamination check, mainly because most plate is manufactured using a vacuum degassing method which reduces significantly the occurrence of laminations. The relevant technical standard is EN 10160 (which replaced BS 5996). In its acceptance criteria it specifies various *body grades* and *edge grades* corresponding to different numbers of imperfections per unit area. The principle is that a plate is tested to see if it complies with the particular grade specified – if it fails, it may be downgraded to the next lower grade. Special high-frequency probes are required for austenitic steels. Material that is 100% checked is often certified and hard-stamped by a third-party organisation for use in statutory pressure vessels and cranes. Figure 7.5 shows points to check when witnessing this technique.

UT of castings

Ferritic and martensitic steel castings can be examined using straightforward ultrasonic methods. The main technique is known as the A-scope (or A-scan) pulse echo method – the same technique is used for

FIG 7.5
Ultrasonic testing of steel plates

Material 'edge'-grades

Acceptance 'grade'	Single imperfection Max length (area)	Multiple imperfections Max no. per 1 m length:
E1	50 mm (1000 mm^2)	5
E2	30 mm (500 mm^2)	4
E3	20 mm (100 mm^2)	3

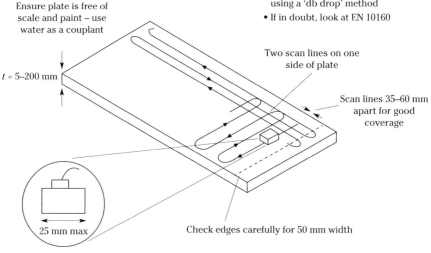

- Use a 'pulse echo' A-scope technique (2–5 MHz)
- Imperfections are delineated by area using a 'db drop' method
- If in doubt, look at EN 10160

Ensure plate is free of scale and paint – use water as a couplant

t = 5–200 mm

Two scan lines on one side of plate

Scan lines 35–60 mm apart for good coverage

Check edges carefully for 50 mm width

25 mm max

0° probe is used

Material 'body' grades

Acceptance 'grade'	Single imperfection Max length (area) (approximate*)	Multiple imperfections Max no. per 1 m length:
S0	5000 mm^2	20
S1	1000 mm^2	15

* these are approximations – if results are marginal, look at EN 10160

forgings. Figure 7.6 shows details. It is a reflection technique (the waves are transmitted and received by the same probe, which has two crystals) and it uses compression waves; each wave oscillates longitudinally along the axis of the beam. The time-scale, which represents distance into the object under test, is always displayed on the horizontal axis. The so-called B-scope method uses a vertical time-base but is in less common use.

Ultrasonic testing of castings is a much simpler technique than that used to examine welds. Its main application is for castings with wall thickness > 30 mm and the techniques are the same whether the casting is designed for general or pressure service. Because of the complex geometry of many castings, full examination by ultrasonic means can sometimes be difficult. Together with the nature of the defects that occur in castings, this leads to a certain amount of freedom being incorporated into the applicable technical standards. They provide guidelines, rather than definitive rules to follow. This is particularly the case with acceptance criteria; codes and standards are oriented towards describing 'grades' of castings, rather than defining absolute accept/reject criteria. Such decisions are left to *agreement* between the manufacturer and purchaser.

As an SI you may need to check the contract specification for any clauses that constitute this agreement; there is often a section entitled 'general requirements for castings', or similar. If there are no clearly specified requirements, you should follow the guidance on technique and defect assessment given in the relevant code or standard, then apply your judgement. As always, *code compliance* is the key point.

UT techniques

The UT examination technique itself is relatively straightforward, but there are a few particular requirements for cast components (see Figure 7.7). When witnessing the test check the following points.

- A 0° (normal) probe is commonly used. It is difficult to use an angle probe technique on castings because of variations in wall thickness of castings (although you may see it specified for large rotating machinery casings).
- The casting *wall thickness* should be divided into zones as shown in Figure 7.7.

The ultrasonic test of a casting follows a well-defined set of steps. As an SI, you need to liaise closely with the test operator to keep track of what is happening. Look for these essential steps.

FIG 7.6
Ultrasonic testing – the A-scan pulse-echo method

Note these points

- A 'pulsed' wave is used – it reflects from the back wall, and any defects
- The location of the defect can be read off the screen

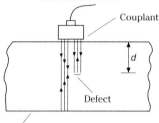

The probe transmits and receives the waves
Couplant
d
Defect
Back wall

The A-scope screen looks like this

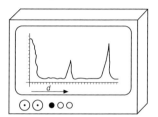

The horizontal axis represents time – i.e. the 'distance' into the material

- *Calibrate* the equipment – first using test blocks and then on the area of greatest wall thickness of the casting itself.
- Do a *preliminary scan* over 100% of the casting surface at a frequency of 2 MHz, changing to smaller probes if necessary to accommodate tight radii. The purpose of this preliminary scan is to locate both planar *and* non-planar discontinuities, but not to investigate them fully at this stage.
- Before proceeding further, *agree the technique* that will be used to 'size' both planar and non-planar discontinuities. They must be sized if the test is to have any real relevance in verifying code compliance of the casting. The most common method is known as the *6 dB drop* technique. This is simply a method by which the operator defines the edge of a defect as that position which causes a 6 dB (50% of screen height) drop in the back wall echo. There is also a more sensitive method termed the *20 dB drop* technique.
- Assess the planar discontinuities for size by finding the location of the edges of each discontinuity.
- Assess the non-planar discontinuities for size. There are two parts to this: finding the upper and lower bounds of the discontinuity, and

FIG 7.7
UT method for castings

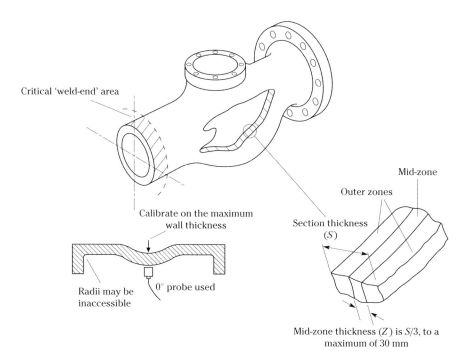

Follow these steps
- Check the suitability of the casting for an ultrasonic technique (the sound permeability)
- Do a visual inspection
- Do a preliminary scan – looking for both types of discontinuity
- Assess the planar discontinuities
- Assess the non-planar discontinuities (find their upper and lower 'bounds' and then delineate their area)

then finding (delineating) the edges, in order to calculate the overall volume.
- Record all the details carefully. It is necessary to describe the sizes and orientation of discontinuities before they can be properly compared with the assessments grades.
- Castings with an austenitic grain structure can be very difficult to test. A general check for sound absorption should be carried out to see if a casting is suitable for being tested ultrasonically. Check with the operator if permeability is > 15 dB. If it is, UT results may be poor and RT is probably better.

Evaluating the UT results

With UT, the purpose of testing castings is to identify *discontinuities*. There are two distinct types: *planar* (in one plane only with no thickness) and *non-planar* (having multiple dimensions and hence an area or volume). A key principle is that planar and non-planar discontinuities are treated *separately* – each type having a different set of criteria by which the casting can be allocated a 'grade'. Note how the sizes of discontinuities are relatively large, compared to the very small defects that are important when examining welds. Castings *will* contain discontinuities and it is not good enough just to refer to 'minor' or 'insignificant' defects in your report. Accurate statements on size are what is needed.

UT of forgings
For forged components that contain thick (> 150 mm) material sections, UT is the only type of volumetric NDE that can be performed. For static equipment it is most commonly used on globe valves and high-pressure thick-walled nozzles and flanges. The NDE procedure for such components normally comprises 100% UT and PT/MT; it is rare to find critical areas defined as is often the case with castings.

Technical standards covering UT of forgings detail the UT techniques themselves rather than defect acceptance criteria. Some do not address acceptance criteria at all, instead referring to the necessity for agreement between manufacturer and purchaser to define what is acceptable. From an SI's viewpoint it can therefore be a difficult task to interpret any discontinuities found in terms of their effect on code compliance and integrity, even though the examination technique itself is relatively straightforward.

Forgings are normally ultrasonically tested twice: before machining (when only a rough assessment is possible) and then after final heat

FIG 7.8
Ultrasonic testing of forgings

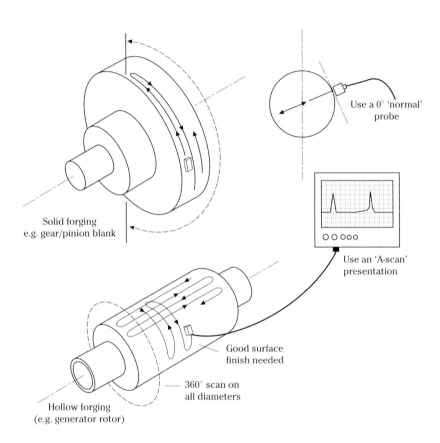

Note these points
- There is no common 'grading system' for discontinuities
- Acceptance criteria are difficult to define; much relies on specific agreements between manufacturer and purchaser

treatment and machining. Figure 7.8 shows the basic techniques that are used for the most common types of forgings that you will see. Note the following points.

- An A-scope presentation is used, as shown in Figure 7.6.
- Some specialised forgings may have an additional specified activity, a *near-surface defect test*. This needs a high-frequency probe (10 MHz) to minimise the thickness of the near-surface 'dead zone' that is a feature of ultrasonic examination.
- The scanning method is usually quite simple on items with constant cross-section. It may be more complicated for components such as valves, which have more complex geometry.
- A normal (0°) probe is used for all preliminary scans. It may be necessary to use angle probes to accurately locate and size discontinuities.
- There are four basic methods of sizing discontinuities; the 20 dB drop and 6 dB drop methods outlined previously cannot be used in all applications and so their use is limited. The *distance-amplitude* and *distance-gain size* techniques have greater application. You can find descriptions of these techniques in published standards and NDE text books.
- It is important for the size of a discontinuity to be accompanied by a description of the technique that has been used to *determine* that size. Sizing techniques are difficult, so it is wise to expect and accept some variability in the results.

UT of welds
Some manufacturers use ultrasonic examination as the exclusive volumetric NDE method for pressure vessel welds. The ultrasonic techniques used are more involved and varied than those used for standard forgings, castings or plate. Compression and transverse (shear) waves are used, the ultrasonic beam being transmitted using a variety of normal and angle probes from several locations on and around the weld. Results are displayed using the A-scope presentation. The technique cannot always be used effectively on austenitic steels unless special high-frequency probes are used, as the grain structure distorts the transverse wave-form.

A large number of weld types, such as butt, nozzle, fillet and branch welds, can be examined using UT. Each weld type has its own 'best technique' for finding defects. From an SI's viewpoint this means that ultrasonic weld examinations present a challenge of *comprehension* – the need to understand a range of techniques if you are to be able to witness such tests competently. Help is available in the form of some particularly informative technical standards that address this subject.

Source inspections and UT codes

As an SI you can expect to become involved with NDE codes and standards on an almost daily basis. Some type of NDE is specified on most types of fixed equipment, with NDE of both the bulk construction materials (pipe, plate, forging and so on) and welded fabrications being common. For welded items, UT is very common with many different permutations of techniques, extent and acceptance criteria.

Current trends in UT codes

Although ASME UT codes have remained in much the same format over time, European UT codes have undergone significant changes over recent years. National standards (British standards, German DIN standards and so on) have all been subject to a process of European *harmonisation*. This has had far-reaching effects, resulting in the withdrawal and replacement of well-established national standards with Euronorm (EN) standards. These are more documents of technical *consensus* rather than necessarily technical excellence. In the field of UT standards, the results have manifested themselves as

- *fragmentation* of large standards into multiple parts, each published separately, covering separate areas of the subject
- *more technical alternatives* of ways to do things in preference to a single 'best' method or solution (because consensus doesn't work like that)
- *more code lists*. There are lots of lists of things that 'should be considered' or included in an activity without always adding detail that would encourage decision making. These are hard to beat as uncontroversial outputs of consensus, so you can see them popping up everywhere.

The above points do not necessarily result in a reduction in the technical quality of the documents, but they can make it a little more difficult to get to the real technical experience that lies behind all the code parts, sections and appendices.

Using UT standard sets

Most European UT standards (EN, ISO and so on) now exist to be used as part of a set, rather than in isolation. As an example, to work on inspection of UT of simple welded joints, you could easily have to refer to

- EN 17640: for UT testing techniques
- ISO 23279: for characterisation of indications
- ISO 5817: for acceptance criteria
- EN 473: for qualification of UT personnel.

There are also others covering more specialised topics such as calibration of test equipment and suchlike. Many of these standards have equivalent designations from several standards bodies and institutions (BS/EN/ISO and so on), all making for a forest of classification and abbreviations. Part of the role of a good SI is to learn the way to ease yourself through this complexity, concentrating on the essential technical detail and ignoring all the peripheral procedural and administration stuff that you don't need.

An example: EN 17640: UT techniques

EN 17640 carries the rather grand title; *Non-destructive testing of welds – Ultrasonic testing techniques, testing levels and assessments.* In short, it describes various UT scanning techniques that are effective in finding indications (and defects) that may exist in a welded seam. All the other things, such as what the indications are called (characterisation) and how relevant they are to integrity, are just cross-referenced to a list of other standards. EN 17640 is a useful operational document for SIs so we will look at some of its important content. The same principles extend across other UT codes such as ASME V and those from other standards bodies.

Figure 7.9 shows the content of the standard diagrammatically. All the parts contribute together to make the UT technique that is effective at finding defects present in the welds and their surrounding material. As an SI it is your role to check that UT procedures carried out are compliant with the relevant standard. Note the main points.

- **Probe size**
  Probe size is governed by the size of the crystal within the probe (called the element). As explained in EN 17640 (6.3.3), they range in size from about 6 mm to 24 mm, governed mainly by the range of the

ultrasonic beam path. Small probes are more suitable for short beam path, with angle probe size above 12 mm required when the required length of the beam paths rises above about 200 mm. Small probes are, however, better at *sizing* defects, once they have been found.

- **Acceptable air gaps**

    UT becomes unreliable when either the shape or the 'waviness' of the component surface results in air gaps beneath the probe. Curved pipe surfaces are a good example. The result is that some of the component volume remains unscanned, so more scans are required to achieve an acceptable scan volume. A typical requirement (from EN 17640 (6.3.4)) is

    Maximum acceptable gap (*g*) in mm between the probe and component surface

    $$g \leq \frac{a^2}{d}$$

    where *a* (mm) is the probe dimension in the direction of testing and *d* (mm) is the diameter of the component surface (e.g. pipe). If $g > 0.5$ mm, then a shaped probe is normally used to reduce the gap.

- **Scanning volume**

    UT technique codes specify the volume of the material that must be scanned in order to adequately find any defects that exist in or near the weld. Figure 7.10 shows this – note how the area extends to the 'full skip' distance plus a bit extra. This area must be checked for laminar defects with a 0° (compression) probe before checking with the angle (shear wave) probe. Laminar defects in this area will cause spurious angle probe indications, which may hide actual defects in the weld itself.

- **Scanning angles**

    Shear wave angle probes use various angles, between the 'critical angle' limits of 35° and 70° (see Figure 7.10). The idea of angle probe examination is to use the angle that best achieves the objective of the beam being incident at 90° to the weld fusion face. This is the most likely place for lack of fusion and slag defects, and so on, so gives the best chance of finding them. An important SI role is to check that the required minimum number of scan angles has been used – several may be required, depending on material thickness and the level of assessment that has been specified. EN 17640 Annex A shows the requirements for several weld configurations (butt, nozzle, T-joint etc. welds).

- **Testing level**
  EN 17640 makes provision for three different levels of testing: A to C. These are based mainly on the extent of testing coverage; that is, the number of scans that are carried out to look for defects. The testing levels are linked to the related code EN 25817 *Quality levels for imperfections* – this provides for three different levels of defect acceptance, B, C and D. Taken together, this system gives three procedure options, with increasing level of probability of detection (PoD) and evaluation of integrity-threatening weld defects. Figure 7.11 shows the levels.
- **UT sensitivity level**
  You can think of the *sensitivity* level of a UT technique as the *volume setting* of the ultrasonic sound waves. The scale is measured in decibels (dB) and the greater the dB, the larger the reflection obtained from a defect reflection, thereby making it easier to find and characterise. Sensitivity is set up before commencing a UT procedure and checked regularly, as the level can be changed by temperature, probe wear and wandering of the UT machine settings. UT technique codes, such as EN 17640, all contain requirements for setting sensitivity, and the levels to be achieved.

The weld assessment

There is a well-defined routine to follow when performing an ultrasonic test on a weld. The basis is the same for butt, nozzle and fillet welds. See Figure 7.12. The five steps are described below.

- Check the *parent material*. The purpose is to check the areas of parent material that the beam will pass through when the weld itself is being examined and to confirm the material thickness so that the beam paths for the angle probe views can be determined. The examples in Figure 7.12 show how the angle beam paths need to be reflected from the inside surface of the parent material in order to assess the weld itself.
- Check the *weld root*. This is most relevant to butt welds that have a separate root run (often GMAW/MIG). For single-sided welds using a backing strip, the root will be undressed and must be examined for root cracks and incomplete root penetration. The examination should be done twice, once from either side of the weld.
- Check using a *normal* (0°) probe. All butt welds should be scanned over the surface of the weld with a 0° probe to detect any root concavity, lack of fusion into the backing strip or poor inter-run fusion.

FIG 7.9
EN 17640 : UT techniques
– The basic content –

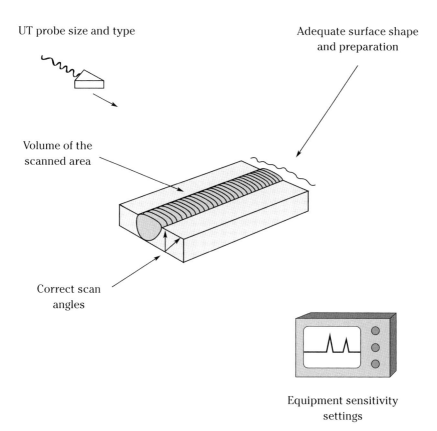

These controlled parameters are common to *all* standards/codes covering UT techniques.

FIG 7.10
EN 17640: UT techniques
– The control of scan patterns –

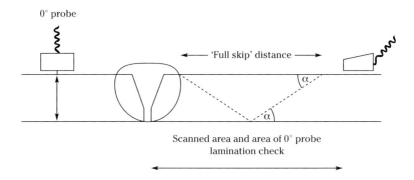

- Probe 'angle' is (α) : between 35° and 70°

- EN 17640 requires minimum of either one or two scan angles depending on the type of weld and assessment level.

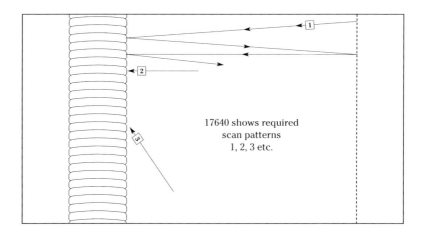

FIG 7.11
EN 17640: testing levels

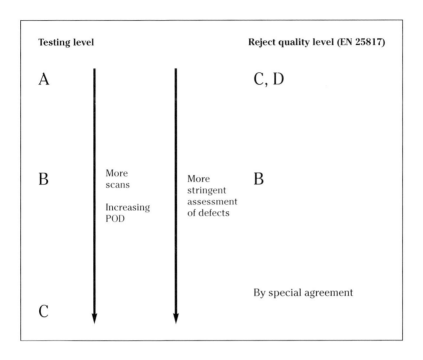

Ref: EN 17660 Table 5: *Recommended testing levels*

Make sure that
- the weld cap is ground smooth enough to allow a proper examination
- probe frequency is a minimum of 4 MHz
- for thick material (say > 50 mm) separate scans are done from the upper and lower surfaces of the weld.
- Scan for longitudinal defects. Here, the beam angles and scanning direction are transverse (across the weld) in order to detect any defects that lie longitudinally along the weld axis. It is common to use several different angle probes; the best chance of detecting a defect is when the beam hits the surface of the defect at 90°. Watch for cracks or lack of fusion lying along the fusion faces.
- Look for transverse defects. These are found by scanning along the axis of the weld. Note the following guidelines.
  - The weld cap should be ground reasonably smooth.
  - Probe angle should be within 20° of the normal. Two probes are needed on welds > 15 mm thickness.
  - Thick welds > 50 mm should be examined from both sides.
  - Corner and T-joints need special scanning techniques. Refer to EN 17640 if you encounter these.

Evaluating indications

The evaluation of indications is a tricky operation. The principles are straightforward but it is the large variety of beam paths and indications that makes it difficult. For this reason it is essential that you work closely with the ultrasonic operator when a defect is found. This is not an area where an SI can afford to 'stand back', relying on a later, more considered analysis of the results; you have to gather your information while the test is being done. The *principles* under which defect assessment operates are, broadly, as follows.

- The assessment is related to the examination level that is being used. For high examinations, any indication above background noise (also called 'grass') needs to be investigated. For less stringent levels, only indications that cause a significant echo are investigated.
- As a rule of thumb, any echo of more than 50% screen height (or a corresponding 50% reduction in back wall echo) merits investigation.
- Defects must be classified (by type), described and measured if they are to be reported with any real meaning. This means that the operator should complete a detailed report sheet when a defect is found. Be careful also of the way in which *you* report defects found –

Inspecting non-destructive examination (NDE)

FIG 7.12
Ultrasonic testing of butt and nozzle welds

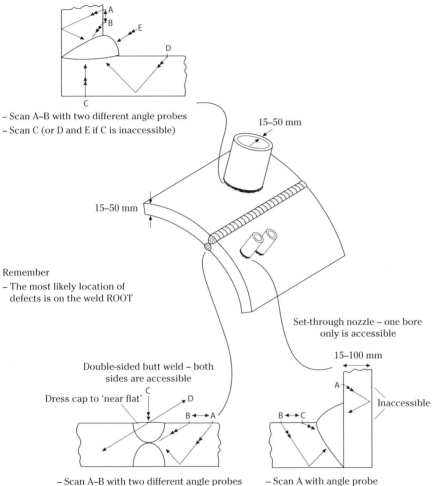

Set-on nozzle with both bores accessible

- Scan A–B with two different angle probes
- Scan C (or D and E if C is inaccessible)

15–50 mm

15–50 mm

Remember
- The most likely location of defects is on the weld ROOT

Double-sided butt weld – both sides are accessible
Dress cap to 'near flat'

- Scan A–B with two different angle probes (for root and longitudinal defects)
- Normal scan C
- Longitudinal scan D with angle probe (for transverse defects)

Set-through nozzle – one bore only is accessible

15–100 mm

Inaccessible

- Scan A with angle probe
- Scan B–C with two different angle probes

you must include full details of the type and size. It is poor practice just to refer to 'defects' without further qualification.

Radiographic testing (RT)

Despite the fact that the *mandatory* requirement for radiographic NDE is generally weakening, you will still find it used extensively in general engineering manufacture. It has the clear advantage of producing a permanent visual record of the results of the examination, enabling the results to be reviewed and endorsed by all of the necessary parties. A key technical advantage of radiography is its ability to identify important types of volumetric indications (porosity, inclusions, lack of fusion and lack of penetration, among others) and categorise them in terms of size. This makes RT very suitable for finding those common types of indications found in multi-layer welds.

A lot of the technology and standards of RT come from pressure vessel practice. Hence you will find that the technical aspects of RT examination are very well covered by the various technical standards. With acceptance criteria, as discussed earlier, some interpretation and judgement is inevitably required.

In this section we shall look at a typical RT application, a full penetration butt weld and nozzle welds in steel pipework. Such pipework would be used for high-temperature application under the ASME code or one of the equivalent European standards. An effective source inspection consists of a well-defined series of steps. We shall follow these through in order.

Check the technique

Figure 7.13 shows the position of the four welds to be examined. Weld no. 1, the large-bore pipe butt weld, can be accessed from both sides. It can be easily radiographed using a single-wall technique. Weld no. 2 is a small-bore full-penetration pipe butt weld. It can only be easily accessed from the outside, so it needs to be examined using a 'double-wall' method. In this case a double-wall, double-image technique is shown. Weld no. 3 is a nozzle weld of the set-through type, as the small tube projects fully through the pipe wall. This weld could be radiographed, but the technique would be difficult, mainly owing to the problem of finding a good location for the film. Attempts to bend the film around the outside of the small-diameter nozzle weld would cause the film to crease, giving a poor image. As a general principle, nozzle to shell welds are more suited to being tested by UT; RT will be difficult and impractical. Weld no. 4, the pipe-to-flange joint, is a *weld neck* flange

FIG 7.13
High-pressure pipework
– welds for volumetric NDT –

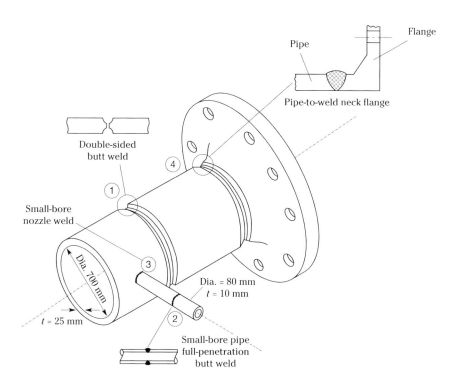

joint. This is a single-sided weld with a root land so qualifies as a full-penetration weld and would require full NDE under most specifications and codes. This joint is capable of being radiographed using a single-wall technique similar to that shown for weld no. 1, as long as there is sufficient room for the film, that is, the width of the weld that can be examined should not be restricted by the position of the flange. In most cases you will find that there will be sufficient clearance. It would also be possible to use UT on this weld if required, scanning from the pipe outer surface and flange face.

Before reviewing the results of any RT examination it is wise to look at the details of how the examination was performed, that is, the examination *procedure*. This may be a separate document, or included as part of the results sheet. Figures 7.14 and 7.15 show examples for welds no. 1 and no. 2. Check the following points.

Single- or double-wall technique? Weld no. 1 is a single-wall technique – the film shows a single weld image shot through a single weld thickness. Ten films are required to view this weld around its entire circumference. In contrast, weld no. 2 is a double-wall technique, suitable for smaller pipes. The offset double-image technique shown is used to avoid the two weld images being superimposed on the same area of film, and hence being difficult to interpret. Standards such as EN 1435 *RT examination of welds* provide useful details.

- Position of the films. Note for weld no. 1 how the ten films, each approximately 230 mm in length, locate around the circumference. Starting from a datum their position is indicated by location marker numbers, so the position of any defects seen in the radiograph can be identified on the weld itself.
- The type of source used. You only need do a broad check on the suitability of the source *type*. For all practical purposes this will be either X-ray or gamma-ray.
- X-rays produced by an X-ray tube are only effective on steel up to a material thickness of approximately 150 mm. Check the voltage level of the source unit: weld thickness of up to 10 mm requires about 140 kV. Thicknesses in excess of 50 mm need about 400–500 kV. A practical maximum is 1200 kV.
- Gamma-rays, produced by a radioactive isotope, can be used on similar thicknesses, but definition is reduced. If a cobalt 60 gamma-source is used, the best results are only obtained for material thickness between 50 and 150 mm. It does not give good results on thin-walled tube welds.
- It is not possible to compare accurately results obtained by X-ray and gamma-ray methods.

Check the films

Although 'real-time' viewing methods are technically feasible, most source inspection techniques use a photographic film which is developed after exposure. A visual record is therefore provided for review by all parties concerned. To check RT films properly requires various essential steps, as described below.

- Check the film *location*. Make sure you relate each film to its physical location on the examined weld or component. Use the procedure

FIG 7.14
RT technique for large-bore butt weld (no. 1)

A single-wall X-ray technique is used:
- the large bore allows access
- the source-outside technique is commonly used for large pipe

The technique looks like this:

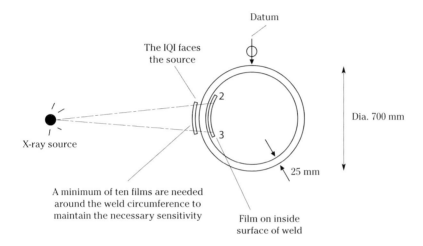

sketch and position markers but *also* look for recognisable features such as weld-tees, flanges and surface marks to perform a double check.

- Check *sensitivity*. This is a check to determine whether the radiographic technique used is sensitive enough to enable indications to be identified if they exist. It is expressed as a *percentage* – a lower percentage sensitivity indicates a better, more sensitive technique. A typical quoted sensitivity is 2%, indicating that, in principle, the technique will show indications with a minimum size of 2% of the thickness of material being examined.

The actual value of sensitivity is determined using a penetrameter, also called an image quality indicator (IQI). The general principle is that one of the IQI wire diameters is a given percentage of the material thickness being examined. If this wire is visible, it shows that the RT technique is being properly applied. There are several commonly used types that you will see; they are described by EN 462 and ASME/ASTM E142 standards. Figures 7.16 and 7.17 show two types and how each is used

FIG 7.15
RT technique for small-bore butt weld (no. 2)

A double-wall, double-image technique is used.

The technique looks like this:

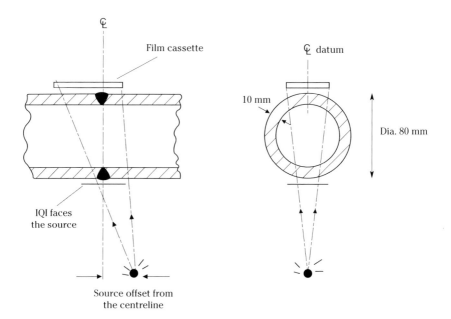

to calculate the sensitivity. Note that you cannot make an accurate comparison between different types – it is essential therefore to quote a sensitivity value against a specific IQI type for it to be capable of proper verification. If sensitivity is worse than that specified then the technique is unacceptable, as it is not capable of finding relevant defects. From an SI's viewpoint, there is normally little point in proceeding further. Issue an NCR at this point, rather than trying to conclude *why* the sensitivity is poor. This is a complicated issue involving all of the technique parameters, so leave it to the specialists.

- Check *unsharpness*; this is properly termed *geometric unsharpness*. This is a measure of the contrast of the image – the difference between the dark and light areas. Although methods have been

Inspecting non-destructive examination (NDE)

FIG 7.16
Using the wire-type penetrameter

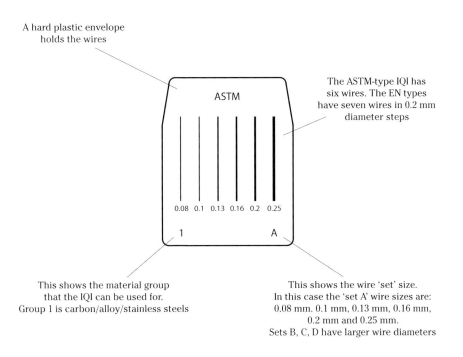

The European standard for IQIs is BS EN 462

- A hard plastic envelope holds the wires
- The ASTM-type IQI has six wires. The EN types have seven wires in 0.2 mm diameter steps
- This shows the material group that the IQI can be used for. Group 1 is carbon/alloy/stainless steels
- This shows the wire 'set' size. In this case the 'set A' wire sizes are: 0.08 mm, 0.1 mm, 0.13 mm, 0.16 mm, 0.2 mm and 0.25 mm. Sets B, C, D have larger wire diameters

- The objective is to look for the smallest wire visible
- Sensitivity = diameter of smallest wire visible/maximum thickness of weld
- If the above IQI is used on 10 mm material and the 0.16 mm wire is visible, then sensitivity = 0.16/10 = 1.6%
- Check the RT standard for the maximum allowable sensitivity for the technique/application being used

devised to measure the degree of contrast, it still involves subjectivity. Check that the image is sharp and defined, without any obvious blurring – you can get a reasonable impression by looking at the IQI. If you do find problems with unsharpness, check the technique sheet for details of the source-to-film and object-to-film distances used; incorrect distances are one of the common causes of unsharpness.
- Check the film *density*. You can think of density as the 'degree of

FIG 7.17
How to read the ASTM penetrameter (IQI)

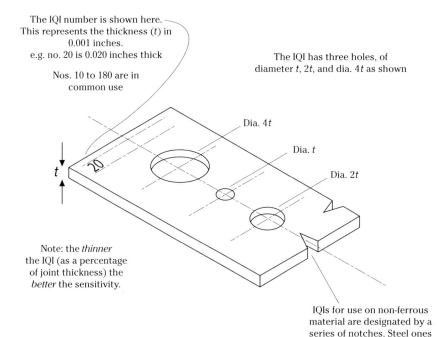

The IQI number is shown here. This represents the thickness (t) in 0.001 inches.
e.g. no. 20 is 0.020 inches thick

Nos. 10 to 180 are in common use

The IQI has three holes, of diameter t, $2t$, and dia. $4t$ as shown

Dia. $4t$
Dia. t
Dia. $2t$

Note: the *thinner* the IQI (as a percentage of joint thickness) the *better* the sensitivity.

IQIs for use on non-ferrous material are designated by a series of notches. Steel ones have no notches.

Image quality designation is expressed as

(X)–$(Y)t$:
(X) is the IQI thickness (t) expressed as a percentage of the joint thickness
(Y) (t) is the hole that must be visible

IQI designation	Sensitivity	Visible hole*
1–2t	1	2t
2–1t	1.4	1t
2–2t	2.0	2t
2–4t	2.8	4t
4–2t	4.0	2t

* The hole must be visible in order to ensure the sensitivity level shown

blackness' of the image. It is determined by an instrument known as a densitometer. Typical values should be between 1.8 and 4.
- *Now* view each film for *indications*. The principle is that indications are identified and then classified according to type and size. Indication *classifications* are well accepted and defined so there is much less room for interpretation than with, for instance, the acceptance criteria for defects, once classified. Published standards contain reference radiographs showing what the various indications look like, for both X-ray and gamma-ray techniques. In practical source inspection situations, such reference radiographs are used less often than you might expect – a simpler list of generic indication types is normally used.

Marking up the films
This is a key role for the SI. Work through the test report sheet viewing each RT film in turn, checking the reported indications carefully against what you see on the films. Mark important indications on the film with a china-graph pencil, using the correct abbreviations. If a repair is required, mark the film 'R' in the top right-hand corner, making a corresponding annotation on the report sheet. Keep each packet of films and its report sheet together and check they are correctly cross-referenced in case they become separated.

There are a few procedural guidelines which you may find useful when reviewing RT results as an SI.

- Before you start checking the films *ask* the manufacturer to summarise the results. Ask whether they are all acceptable and which acceptance criteria have been applied. Wait for the answer before you pick up the first film.
- Check the films showing indications *first*. Start with the worst ones, that is those exhibiting the major indications. Do not waste time discussing minor film or surface marks.
- Be aware of the acceptance criteria specified in the common pressure vessel codes. Surprising amounts of defects such as porosity *are* allowed – the amount depends on the particular criteria being applied. You cannot normally demand absolutely clear radiographs.
- It is best if any defects have already been repaired – but ask to look at both 'before' and 'after' radiographs to satisfy yourself what has been done.
- Remember the principle explained in an earlier chapter that it is the SI's job to guide a *consensus on code compliance*. This is an important point in technical areas like this where interpretation and judgement

FIG 7.18
Watch out for welding and NDE non-conformances

1. Welding and NDT are separate disciplines but for inspection purposes, you can consider them as being closely *linked*.
2. The key criteria are *code compliance* and weld *integrity*. The role of a source inspector is to *verify* this integrity, not to participate in detailed discussions about techniques.
3. Be prepared to use your judgement when considering acceptance criteria – the situation may *not* always be absolutely clear.
4. Weld indications and defects have predictable causes and effects.
5. Full penetration welds need the most stringent NDE. This is due to the high probability of weld defects being located in the weld root, which cannot be ground (back gouged) if access is restricted.
6. Welding 'controlling' documentation (WPSs, PQRs and welder approvals) is important but it is not an end in itself. Don't concentrate too much on documentation instead of inspecting the weld itself.
7. Surface crack detection. PT and MT techniques really only provide an enhanced visual assessment.
8. Volumetric NDE. Many equipment standards accept *either* RT or UT techniques; they do not find it easy to express a preference.
9. Try to learn a little about the various UT techniques (particularly the examination of welds) so you can understand what is happening when you witness a test.
10. Normally a source inspector *cannot* demand absolutely 'clear' radiographs – you have to use the agreed acceptance criteria.
11. Destructive tests consist of tensile, bend and macro examination/hardness tests across the weld section. Learn to describe accurately the results of these tests.

are *always* part of the equation. Do not neglect the experience of any third-party inspector present if you are reviewing weld radiographs relating to a statutory equipment item.

Chapter 8

Inspecting pressure vessels

The inspection of vessels and other pressure equipment forms a major part of an SI's role. Any chemical or process plant will have a large number of vessels for different applications. Some are complex, such as those forming the component parts of steam-raising plant or large condensers; others such as air receivers and low-pressure or atmospheric vessels are of relatively simple design and construction. The scope is wide but all present the SI with a similar task. In this chapter we shall look at the general principles and technical issues of source inspection, drawing together relevant areas from previous chapters dealing with codes and standards, materials of construction and NDE. Vessels provide a useful vehicle for showing how different inspection disciplines mesh together in a practical source inspection context. We shall also look at the implications of the fact that many pressure vessels are subject to statutory certification requirements.

8.1 Compliance and integrity criteria

Pressure vessels can be dangerous. In use, they contain large amounts of stored energy, so design and construction codes have been developed to give a reasonable confidence that the vessel is safe and is not going to fail unexpectedly. Figure 8.1 shows four core subjects incorporated into all pressure equipment design and construction codes

- independent design appraisal
- traceability of materials
- NDE
- pressure testing.

Your role as an SI is to verify that all of these elements are in place and have been completed, thereby making you a part of the *control mechanism* for the equipment construction. In a real source inspection situation there are often several parties involved. The manufacturer,

FIG 8.1
The pressure vessel 'norms'

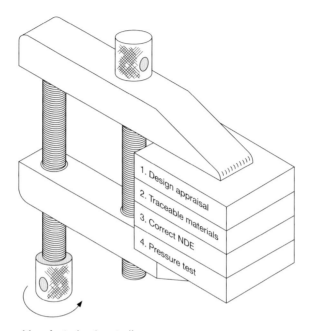

Manufacturing 'control'

Vessel integrity (and code compliance) is obtained by:
- Arranging for an independent design appraisal
- Using traceable materials
- Applying proven NDT techniques
- Doing a hydrostatic (pressure) test

and then

- Exerting proper (*meaning enough*) control over the manufacturing process

main contractor and statutory inspector all have an *interest* in code compliance, separated by a bit of a difference in focus. Try to rationalise things by accepting that this is an inherent part of the system of norms that has developed over time. If you are involved in the inspection of pressure vessels, you will very soon meet the issue of *statutory certification*. We shall look now at how this forms a part of the concepts of compliance and integrity.

8.2 Statutory certification

There are a number of common misunderstandings surrounding the statutory certification of pressure vessels so do not be surprised if you occasionally find some confusion, even among experienced companies and their staff, as to what *certification* means and implies. It is useful therefore to have an understanding of the key issues so you can act effectively during a source inspection.

Why is certification needed?

There are four possible reasons why a pressure vessel needs to be certificated, as follows.

- The need for certification is imposed or inferred by statutory legislation in the country where the vessel will be *installed and used*.
- The need for certification is imposed or inferred by statutory legislation in the country where the vessel is *manufactured*.
- The need for certification is imposed or inferred by the company that will provide an insurance policy for the vessel itself and second- and third-party liabilities when it is in use.
- The manufacturer, contractor or end-user *chooses* to obtain certification because they feel that
  - it helps to maintain a good standard of design and workmanship
  - it provides evidence to help show that legal requirements for due diligence and duty of care have been met.

Note that three of these reasons are as a result of the certification requirement being imposed (or at least *inferred*) by an external player, whereas the other is a voluntary decision by one or more of the directly involved parties. Perhaps surprisingly, this voluntary route accounts for more than 30% of vessels that are certified. Of the other 70%, which are subject to certification because of the requirements of other parties, it is safe to say that some of these perceived requirements are undoubtedly imaginary rather than real. This is because there are many countries in the world where statutory requirements are unclear, contradictory or non-existent. The more risk-averse vessel manufacturers and contractors often assume that certification *will* be necessary, even if evidence of this requirement is difficult to find.

What is certification?

Probably certification is not precisely what you think. Certification is an *attempt* to assure the integrity (our term) in a way that is generally accepted by external parties. It uses accepted vessel standards or codes such as EN, ASME and so on as benchmarks of acceptability and good practice. Certification does address issues of vessel design, manufacture and testing, but only insofar as these aspects are imposed explicitly by the relevant construction code, no more. Certification is evidence therefore of *code compliance*. Compliance with the ASME code is a special case, as it can be a statutory requirement in some states of the USA. Similarly, pressure equipment to be used in EU countries has to comply with the European pressure equipment directive (PED). This is not a construction code, as such, but does specify a list of essential safety requirements (ESRs) as well as inspection input from a notified body (NoBo) and requirements for declaration of conformity certificates and nameplate/stamping.

So:
If a pressure vessel is fully certified.

Then:
There is hard evidence that the vessel complies with the requirements of the design code or standard stated on its certificate.

In order to obtain full certification, the organisation intending to issue the certificate must comply with the activities raised by the relevant construction code. These differ slightly between codes but the basic requirements are the need to

- perform a detailed design appraisal
- ensure the traceability of the materials of construction
- perform NDE activities and review the results
- do a pressure test
- monitor the manufacturing process in some way
- issue a certificate, or manufacturer's data report (MDR), or PED declaration of conformity in the form required by the relevant code or national legislation.

To avoid misconceptions it is useful to look at what certification is **not**. Vessel certification is normally

- **not** a *guarantee* of integrity (there can be no such thing)
- **not** a statement of fitness for purpose
- **nothing to do** with project-specific engineering aspects of the vessel

such as the position of nozzles, dimensional accuracy, corrosion resistance, mounting arrangements, instrumentation or external painting and internal protection (including shotblasting and preparation)
- **not** a mechanism for the manufacturer to off-load contractual or product liability responsibilities on to the certifying organisation (who will rarely actually *approve* drawings or documents – they will simply be stamped 'reviewed'). Similarly, detailed technical aspects such as concessions are more likely to be 'noted' rather than 'agreed' as part of the certification process.

Often it is these limitations of the certification exercise that are misunderstood, rather than the objectives of the inspections that are carried out by the certifying organisation.

Who can certificate vessels?

There are two aspects to this, *independence* and *competence*.

Independence
The main pressure vessel codes require that vessels, if they are to comply fully with the code, are certificated by an organisation that is independent of the manufacturer. This is generally taken to mean that there should be no direct links in the organisations that would cause commercial (or other) pressures to be imposed on the objectivity of the certificating organisation's actions.

Competence
In European countries (this can be where the vessel is to be used) there exists a restricted list of NoBo organisations that are deemed competent to perform the certification of vessels to the PED. By definition these organisation have been assessed, usually by governmental authorities, for both independence and competence. In the USA, ASME code equipment can only be certified by a National Board commissioned authorised inspector (AI).

In some countries you will simply meet the statement that vessel certification has to be performed by a '*competent*' person or organisation. Do not expect to see convincing guidance as to what qualifies a person or organisation as being competent. Competent status will normally only be challenged if there is an accident, failure incident or negligence claim, when the onus will be on the person or organisation claiming the status to prove it. There are various registration schemes in existence by which organisations may try to improve their perceived status as a 'competent body'; most of these sound convincing, but are voluntary. The majority of such schemes involve the organisation submitting to audits to an international standard with additional requirements added. Of those organisations that do provide certification services for pressure vessels, some subscribe to such schemes because they see some benefit or just think it sounds good. Others do not, as they do not believe they are worth the trouble and cost.

In the eyes of the technical standards that just require certification (or *inspection and survey* is the term sometimes used), all organisations that can meet the independence criteria have the same status. This means that any of the following organisations can do it

- a classification society
- an insurance company
- an independent inspection company
- a government department
- (in theory) an uninvolved manufacturer or contractor.

ASME and the European PED are special cases with their own more restrictive requirements.

There are no hard and fast rules which state that all parts of the certification process have to be performed by the same organisation. You may find in practice that the design appraisal part is separate. In some industries it is traditionally done by a 'design institute' or indigenous classification society, and the source inspection part is then subcontracted to a separate inspection company. This is perfectly acceptable, as long as the certifying organisation is able to accept the validity of the design appraisal performed by the other party.

8.3 Working to pressure equipment codes

It is difficult to find a contract specification for an industrial plant that does not refer to one of the series of internationally recognised technical standards for pressure vessels. These are more generally referred to as pressure vessel *codes*. One of the core tasks of source inspection is to work within the requirements of these codes or, more precisely, *inspect to* a specific code. The main problem is the size and sheer complexity of the code documents. Even with electronic copies this can be difficult, as each code contains an extensive list of cross-referenced technical standards.

The answer is selectivity. As an SI you need only be concerned with those parts of a vessel code that have an impact on the activities of source inspection. Look carefully at the vessel codes and try to obtain an overall insight into the structure of the documents. Although there are some differences between the American (ASME) and European based (EN, ISO and so on) codes, both groups follow similar principles. For the design of the pressure vessel, most of the detailed information is incorporated in the code document itself, whereas for the inspection aspects (and to a lesser degree the testing requirements) the main purpose of the code document is to cross-reference applicable subsidiary technical standards. The result is that vessel codes have a strong tendency to be design orientated (rather, perhaps, than *source inspection orientated*).

This means that to inspect a vessel to a pressure vessel code you only need a limited, and therefore manageable, amount of information from that code. You then also need a selective amount of information from subsidiary technical standards. The best place to start is with Figure 8.2. Look carefully at the content. It lists *only* those pieces of information that an SI needs from a vessel code and the subjects coincide broadly with the entries that you will find in the code index.

Effective source inspection of vessels means letting the manufacturer do some of the work. It may be acceptable to ask the manufacturer to identify a relevant code section or clause and then to demonstrate their vessel's compliance with it. Some will respond and others may not. It is not wise, however, to take this to extremes – you should know the basic material, NDE, acceptance criteria and inspection requirements to enable you to ask the right questions.

Applications of pressure vessel codes

Most of the international pressure vessel codes have been developed to the point where there is some cross-recognition between countries' statutory authorities of each other's codes. Core areas such as vessel classes, design criteria and requirements for independent inspection and certification are based on similar guiding principles. Increasingly, vessel codes are also being adopted for use on other types of engineering components and equipment. This is most evident with Euronorm (EN) and ASME, in which their parameters for allowable stresses and factors of safety are used for guidance in the design of other equipment. This generally involves equipment items having thick cast sections, such as rotating equipment casings and large cast valve chests, or those that have similar construction to pressure vessels, such as high-pressure tubed heaters and condensers. In applications where pressure loading is

FIG 8.2
Information you need from a pressure vessel code

1. Responsibilities

First, you need information on the way that the code allocates *responsibilities* including:

- Relative responsibilities of the manufacturer and the purchaser.
- The recognised role of the independent inspection organisation.
- The technical requirements and options that can be agreed between manufacturer and purchaser.
- The way in which the certificate of *code compliance* is issued – and who takes responsibility for it.

2. Vessel Design

Important points are:

- Details of the different construction *categories* addressed by the vessel code.
- Knowledge of the different classifications of welded joints.
- Knowledge of prohibited design features (mainly welded joints).

3. Materials of Construction

You need the following information about materials for pressurised components of the vessel:

- Materials that are *referenced* directly by the code are subdivided into plate, forged parts, bar sections, and tubes.
- The code's requirement for other (non-specified) materials, in order that they may be suitable for use as *permissible* materials.
- Any generic code requirements on material properties such as carbon content, UTS or impact value.
- Specific requirements for low-temperature applications.

4. Manufacture, Inspection and Testing

The relevant areas (in approximate order of use) are:

- Requirements for material identification and traceability.
- NDE of parent material.
- Assembly tolerances (misalignment and circularity).
- General requirements for welded joints.
- Welder approvals.
- Production test plates.
- The extent of NDE on welded joints/Acceptable NDE techniques.
- Defect acceptance criteria.
- Pressure testing.

· Content of the vessel's documentation package.

not such a major issue you will still find the inspection and testing parts of vessel codes being used. The welding and NDE requirements of both EN codes and ASME VIII/ASME V are frequently specified for use in gas dampers and ductwork, structural steelwork, crane steelwork and similar fabricated equipment.

Vessel codes as 'an intent'

The use of some of the content of vessel codes for other equipment can make the activities of an SI a bit more difficult. You can foresee the problem – an item such as a large condenser or heat exchanger may be specified as being *to ASME intent*. From a design stress point of view this is fine – the problem comes when you try to apply the inspection and testing requirements of the code. Items such as material traceability, the amount of NDE and defect acceptance criteria often just do not fit. This leaves you with the task of inspecting against *partial code compliance*. You will meet this situation quite often in source inspections. Your main task is to rationalise the way and the extent to which the elements of the vessel codes contribute to the integrity of the equipment. Try the following guidelines.

- Look for those component *parts* of the piece of equipment that are the same as those used in the vessel code (see Figure 8.3 for a typical example). Treat these parts as you would a full code pressure component, applying the code requirements just as rigidly.
- Where a component does not match the type of component mentioned in the code (i.e. the component for which the code was originally intended), you need to apply judgement, and be prepared to make a decision based on your engineering knowledge. Look closely at the integrity criteria for the component in question then try to apply as many of the code requirements as you can *without* there being a direct technical contradiction. For example, there is no reason why the ASME code requirement for hydrostatic test pressure cannot be applied to a fabricated intake chest for a sea-water pumping system. Subsidiary standards, particularly those relating to materials testing and NDE, can similarly be applied without any technical contradictions to many equipment types. If in doubt, think again about the integrity criteria for the equipment, then use your judgement.
- A golden rule. Treat the technical requirements of the vessel codes as good and proven engineering practice. Do not be persuaded that they are too specialised to be applied to other equipment. Be careful of

FIG 8.3
Heat exchanger to 'ASME VIII intent'

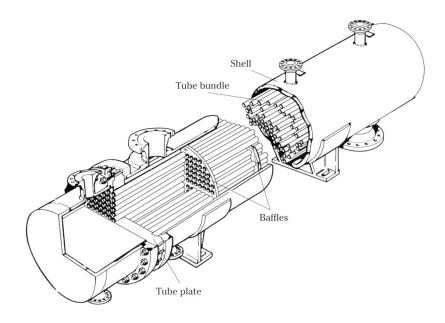

Materials (particularly the internals) may not be those specified by ASME II

Stress calculations do use the assumptions of ASME VIII (but often only for the outer shell – not the tubes)

Tube–tube plate components may be based on Tubular Exchanger Manufacturer's Association (TEMA) (rather than ASME) requirements

The vessel is not ASME-stamped

NDE extent is not governed by the ASME joint types ABCD. A simpler system is often used.

NDE techniques and defect acceptance criteria may be ASME-specified or modified by purchaser/manufacturer agreement.

This figure shows the typical characteristics of an ASME-intent vessel. There are many possible variations from the ASME code – these are the common ones.

over-specified material traceability requirements though; most non-statutory equipment will not exhibit full material traceability. Partial traceability using predominantly EN 10 204 type 3.1 certificates is common (and accepted) in most industries.

8.4 Inspection and test plans (ITPs)

The use of ITPs is well accepted in pressure vessel manufacturing and is one of the most useful *working* documents for the SI. Used in a key monitoring and control role, ITPs summarise the activities of manufacturer, contractor and statutory certification organisation. Owing partly to the statutory nature of pressure vessels, you can expect ITPs to have a well-defined set of technical steps. This tends not to be the case with SI *witnessing* responsibilities, however, as the number and extent of witness points can vary significantly between contracts. In general, it is the contractual agreement between the parties that defines the activities to be witnessed during manufacture. Remember also that the extent of SI mandatory inspection required for statutory purposes is not always explicitly defined. Relevant legislation quotes the main vessel codes as examples of what is considered good technical practice, and some of these are clearer than others.

For these reasons the typical vessel ITP shown in Figure 8.4 concentrates on the *technical* steps included in a good ITP rather than trying to define responsibilities for witnessing these activities. Also included is the inventory of typical documentation items relevant to each step of the ITP. Note the following particular guidelines when dealing with pressure vessel ITPs.

- **Code compliance**. A good pressure vessel ITP will refer to the relevant sections of the applicable code for major topics such as welding procedures, production test plates and NDE.
- **Acceptance criteria**. A reference to the defect acceptance criteria to be used should be shown explicitly in the ITP. If it is not, make a point of holding early discussions with the vessel manufacturer about this. Remember that some interpretation and judgement is still required for defect acceptance criteria written in vessel codes.
- **Hold points**. The use of hold points where a manufacturer must stop manufacture until an SI completes an interim works inspection should be treated carefully. Expect manufacturers to paint a gloomy picture of how hold points can delay the manufacturing programme. In practice this *is* often the case. Try to see the manufacturer's viewpoint and limit formal hold-points to the major manufacturing steps, and the final inspection/hydrostatic test. This does not mean

FIG 8.4
Pressure vessels: Typical ITP content

Activities	Relevant Documentation
1. Design appraisal	Certificate from independent organisation that the vessel design complies with the relevant code.
2. Material inspection at works (forgings, casting, plate and tubes)	Identification record/mill certificate (includes chemical analysis)
2.1 Identification/traceability	Witness identification stamps
2.2 Visual / dimensional inspection	Test report (and sketch)
2.3 UT testing of plate	Test certificate
2.4 Mechanical tests	Mechanical test results (including impact)
3. Marking off and transfer of marks	Material cutting record (usually sketches of shell/head plate and forged components)
4. Examination of cut edges	PT/MT record for weld-prepared plate edges
5. Welding procedures	WPS/PQR records
5.1 Approve weld procedures	Welder qualification certificates
5.2 Check welder approvals	Consumable certificate of conformity
5.3 Verify consumables	Location sketch of test plate
5.4 Production test plates	
6. Welding	Check against the WPS
6.1 Check weld preparations	Record sheet
6.2 Check tack welds and alignment of seams/nozzle fit-ups	PT results sheet
6.3 Back chip of first side root weld, PT test for cracks	WPS and visual inspection sheet
6.4 Visual inspection of seam welds	
7. Non-destructive testing before heat treatment	RT or UT test procedure
7.1 RT or UT of longitudinal and circumferential seams	Defect results sheet
7.2 RT or UT of nozzle welds	NDE location sketches
7.3 PT or MT of seam welds	Repair record (and location sketch)
7.4 PT or MT of nozzle welds	
7.5 Defect excavation and repair	
8. Heat treatment/stress relief	Visual/dimensional check sheet
8.1 Visual/dimensional check before HT	HT time/temperature charts
8.2 Heat treatment check (inc. test plates where applicable)	
9. Non-destructive testing after HT	RT or UT test procedures
9.1 RT or UT of longitudinal and circumferential seams	NDT results sheets
9.2 RT or UT of nozzle welds	NDT location sheets
9.3 PT or MT of seam welds	
9.5 PT or MT of attachment welds, lifting lugs and jig fixture locations	
10. Final inspection	Test certificate
10.1 Hydrostatic test	Record sheet
10.2 Visual and dimensional examination	Record sheet
10.3 Check of internals	Record sheet
10.4 Shotblasting/surface preparation	Record of paint thickness and adhesion test results
10.5 Painting	Record of oil type used
10.6 Internal preservation	Copy of vessel nameplate
10.7 Vessel markings	Packing list
10.8 Packing	Full package, including index
10.9 Documentation package	Code-specified form
10.10 Vessel certification	Record of all concessions granted (with technical justification)
10.11 Concession details	

that you have to give up your right to *witness* important manufacturing steps. You can just as easily do it informally, and without disrupting important manufacturing steps. This means that interim visits *are* required as you cannot expect all manufacturers to do your job for you.

- **Documentation review**. Some manufacturers will only start to compile the documentation package for a vessel after that vessel has successfully passed its visual/dimensional inspection and hydrostatic test. Typically, the package is compiled over a period of several weeks after the hydrostatic test, during which time the vessel is shotblasted, painted and packed ready for shipping. Such a system is not very helpful because any problems of missing or incorrect documentation are invariably discovered too late and it is not unknown for SIs to be placed under pressure to release the vessel before all documentation has been properly reviewed. To lessen the effects of this, make sure that you review all the key documentation during the manufacturing programme. Your review should be substantially complete before you visit the works to witness the hydrostatic test and visual/ dimensional check. Expect to have to provide several reminders about early compilation of the documentation package. *Ask to see it* – do not just remind everyone of its importance.

8.5 Pressure testing

Most vessels designed to operate above atmospheric pressure will be subject to a pressure test at the end of manufacture, normally in the manufacturer's works. Witnessing pressure tests is therefore a common source inspection task. It should always be shown as a witness point for all parties concerned on a vessel's ITP and is an integral part of the role of the third-party organisation or authorised inspector when the vessel is subject to statutory certification.

The objectives of a pressure test

The *objectives* of pressure tests are sometimes misunderstood. It is part of the system of verifying the integrity of a vessel but it has its limitations. The stresses imposed on a vessel during a pressure test are effectively static. This means that what they test is the resistance of the vessel only to the principal stress and strain fields, not its resistance to cyclic loadings (that may cause fatigue), creep or the other mechanisms that have been shown to cause vessels to fail. Hence the pressure test is *not* a full test of whether the vessel will fail as a result of being exposed to its working environment. Frankly, the incidence of steel vessels actually failing catastrophically under a works pressure test is small, almost negligible. A pressure test is **not** a 'proving test' for vessels that have not been properly checked for defects (particularly weld defects). It is also not a proving test for vessels where unacceptable defects have been found – so that the vessel can be somehow shown to demonstrate *integrity*, in spite of the defects.

So

- a pressure test is a test for leakage under pressure

and

- resistance to brittle fracture at the test temperature

and

- that is about *all* it is.

The standard hydrostatic test

This is the most common pressure test performed on steel vessels. It is also commonly known as a hydraulic test. It is a routine test used when the vessel material thickness and allowable stresses are well defined and there are no significant unknown factors in the mechanical aspects of the design. For single-enclosure vessels (such as drums, headers and air receivers) a single hydrostatic test is all that is required. For heat-exchange vessels such as heaters, coolers and condensers, a separate test is performed on each 'fluid side' of the vessel. Figure 8.5 shows the guidelines to follow when witnessing a hydrostatic test. All the main pressure vessel codes provide a formula for calculating the minimum test pressure as a multiple of design pressure. For source inspection purposes you can assume that *design pressure* is the same as *maximum allowable working pressure (MAWP)*. You may also find it referred to as *rated pressure* in some standards.

The test pressure multiplier to be used for hydraulic testing varies with the design and construction code. Typical examples are

ASME VIII-I vessels: Test pressure = 1.3 × MAWP × temperature correction ratio

ASME B31.3 pipework: Test pressure = 1.5 × MAWP × temperature correction ratio

These are *minimum* test pressure values. Some codes also provide a limit for maximum test pressure allowed, to ensure that excessive strain or yielding does not occur during the test.

FIG 8.5
Job well done: Hydrostatic tests

Vessel configuration
- The test should be done after any stress relief.
- Vessel components such as flexible pipes, diaphragms and joints that will not stand the pressure test must be removed.
- The ambient temperature must be above 0°C (preferably 15–20°C) *and* above the brittle fracture transition temperature for the vessel material (check the mechanical test data for this).

The test procedure
- Blank off all openings with solid flanges.
- Use the correct nuts and bolts, not G-clamps.
- Two pressure gauges should be used, preferably on independent tapping points.
- It is essential for safety purposes to bleed all the air out. Check that the bleed nozzle is really at the highest point and that the bleed valve is closed off progressively during pumping, until all the air has gone.
- Pumping should be done slowly (using a low-capacity reciprocating pump) so as not to impose dynamic pressure stresses on the vessel.
- Test pressure is stated in EN/ASME or the relevant standard. This will not overstress the vessel (unless it is a very special design case). Test pressure can vary from 1.3 to 1.5 × design pressure (**MAWP**), plus a temperature correction factor.
- Isolate the pump and hold the pressure for a minimum of 30 min.

What to look for
- Leaks can take time to develop. Check particularly around seams and nozzle welds. Dry off any condensation with a compressed air-line, it is possible to miss small leaks if you do not do this. Leaks normally occur from cracks or areas of porosity.
- Watch the gauges for pressure drop. Any visible drop is unacceptable.
- Check for distortion of flange faces etc. by taking careful measurements. You are unlikely to be able to measure any general strain of the vessel as it is too small.
- If in any doubt, ask for the test to be repeated. It will not do any harm.

Pneumatic testing

Pneumatic testing of pressure vessels is a special case testing procedure used when there is a good reason for preferring it to the standard hydrostatic test. Common reasons are listed below.

> **FIG 8.6**
> **Watch out for: Pneumatic tests**
>
> - Some codes require that a design review be carried out to quantify the factors of safety inherent in the vessel design before doing a pneumatic test.
> - ASME VIII (part UW-50) specifies that all welds near openings and all attachment welds should be subject to 100% surface crack detection (PT or MT).
> - It is *good practice* to carry out 100% volumetric NDE and surface crack detection of all welding prior to a pneumatic test – even if the vessel code does not specifically require it.
>
> **The test procedure**
> - The vessel should be in a pit, or surrounded by concrete blast walls.
> - Ambient temperature should be well above the brittle fracture transition temperature.
> - Air can be used but inert gas (such as nitrogen) is better.
> - Pressure should be increased very slowly, in steps of 5–10%, allowing stabilisation between each step.
> - The normal test pressure is 1.1 × MAWP (design).
> - When test pressure is reached, isolate the vessel and watch for pressure drops. Remember that the temperature rise caused by the compression can affect the pressure reading (the gas laws).

- Refrigeration system vessels are frequently pneumatically tested with nitrogen.
- Large or special gas vessels may have an unsupported structure and so are unable to withstand the weight of being filled with water.
- Vessels which are used in critical process applications where the process of even small quantities of water cannot be tolerated.

The test multiplier for pneumatic testing under most codes is 110% MAWP. Under the ASME code there is an additional temperature compensation ratio to be incorporated as well, but this is only used for VIII-I vessels, not B31.3 pipework.

Pneumatic tests are dangerous. Compressed air or gas contains a large amount of stored energy, so in the unlikely event that the vessel does fail, this energy will be released catastrophically. The vessel will effectively explode, with potentially disastrous consequences. For this reason there are a number of well-defined precautionary measures to be taken before carrying out a pneumatic test on a vessel, and various safety aspects to be considered during the testing activity itself. These are shown in Figure 8.6, along with more general guidelines on witnessing a pneumatic test.

Vacuum leak testing

Vacuum leak tests are different to the standard hydrostatic and pneumatic tests previously described. The main applications you will see are for condensers and their associated air ejection plant. This is known as *coarse vacuum equipment*, designed to operate only down to a pressure of about 1 mmHg absolute. Most general power and process engineering vacuum plant falls into this category. There are other industrial and laboratory applications where a much higher fine vacuum is specified, but this is a highly specialised area, outside the common source inspection field.

The objective of a coarse vacuum test is normally as a proving test on the vacuum system rather than just a vessel itself. A typical air ejection system will consist of several tubed condenser vessels and a system of interconnecting ejectors, pipework, valves and instrumentation. The whole unit is often skid-mounted and subject to a vacuum test in its assembled condition. A vacuum test is more searching than a hydrostatic test and will register even the smallest of leaks that would not show during a hydrostatic test, even if a higher test pressure was used. Because of this the purpose of a vacuum leak test is not to try to verify *whether* leakage exists, but to determine the *leak rate* from the system and compare it with a specified acceptance level.

Leak rate and its units

The most common test used is the *isolation and pressure-drop method*. The vessel system is evacuated to the specified coarse vacuum level using a rotary or vapour-type vacuum pump and then isolated. Note the following points.

- The acceptable leak rate will probably be expressed in the form of an allowable pressure rise (p). This has been obtained by the designer from consideration of the leak rate in Torr litres per second.

$$\text{Leak rate} = \frac{dp \times \text{volume of the pressure system}}{\text{time}, t \text{ (seconds)}}$$

The units are Torr litres per second (Torr ls^{-1}). For source inspection

purposes you can consider 1 Torr as being effectively equal to 1 mmHg. Note also that when discussing vacuum, pressures are traditionally expressed in absolute terms, so a vacuum of 759 mmHg below atmospheric is shown as +1 mmHg.
- It is also acceptable to express leak rate in other units (such as '1 µmHg s^{-1} colloquially known as a *lusec*) and other combinations. These are mainly used for fine vacuum systems. If you do meet a unit that is not immediately recognisable, it is easiest to convert it back to Torr ls^{-1} via the SI system.
- Because leak rate is a function of volume, the volumes of all the system components: vessels, pipes, traps, bypasses and valves need to be calculated accurately. It is not sufficient just to use the approximate design volume of the vessels.

The vacuum test procedure

The procedure for the isolation and pressure drop vacuum test is simple: evacuate the system, close the valve and then monitor pressure rise over time. The main effort, however, needs to be directed towards the

---

FIG 8.7
Watch out for: Vacuum leak tests

PREPARATION IS IMPORTANT
1. Do a standard hydrostatic test on the installation before doing the vacuum test, to identify any major leaks.
2. Leak-test small components (pipes, valve assemblies and instrument branches) before assembly. Submerging the pressurised component in a water bath is the best way.
3. The inside surfaces of all components must be totally clean and dry. Even small amounts of moisture, porous material or grease will absorb air and release it during the test, giving erroneous readings (virtual leaks).
4. Visually check and clean all flange faces before assembly. Polish out radial scratches. Use new pipe joints. Strictly, liquid or paste jointing compound should not be used as it can mask leaks.
5. Do the test under dry, preferably warm (10°C minimum) ambient conditions.
6. The configuration should be such that various sections of the system can be isolated from each other. This helps to locate the position of leaks.

preparation for the test – it is surprisingly easy to waste time obtaining meaningless results if the preparation is not done properly. This happens quite frequently. Figure 8.7 shows points to check before carrying out a vacuum leak test. It is wise not to be too hasty in issuing an NCR for excessive leakage until you have implemented these points. This is more to save your time and effort by not having to make abortive repeat works visits to witness poor tests, than to give the equipment manufacturer the benefit of the technical doubt.

Leaks can be difficult to locate. If the observed leak rate is above the specified levels but still relatively small, make a double check on the tightness of the pressure gauge and instrumentation fittings. These are a common source of air ingress, particularly if they are well used and have worn union connections. The next step is to isolate the various parts of the system from each other to identify the leaking area. The system can then be pressurised using low-pressure air and a soap/water mixture brushed onto suspect areas. Concentrate on joints and connections; leaks will show up as bubbles on the surface.

8.6 Visual and dimensional examination of vessels

The visual and dimensional examination is part of the final source inspection activities carried out on a pressure vessel. Final inspection is mandatory for vessels that are subject to statutory certification, as well as being a normal contractual witness point. It is not a difficult exercise, but it does benefit from a structured approach and the use of checklists to aid reporting. The visual examination and dimensional check can be done before or after the standard hydrostatic test. It is perhaps more common to do it afterwards – this enables internal and external examinations to be done during the same visit. At this stage the vessel is awaiting shotblasting and painting.

The visual examination

The purpose of the visual examination is to look for problems of non-compliance with the code and general arrangement drawings. It is also an important way in which to gain clues about any poor manufacturing practices that may have been used during those manufacturing activities that were not witnessed by an SI.

The vessel exterior
The basic examination principles are similar for all vessels. Check the following points.

- **Plate courses**. Check the layout of the plate courses against the original approved *design* drawings (not the ones that you may find on the shop floor next to the vessel). It is not unknown for manufacturers to change the layout of the plate courses to make more economical use of their stock plate. If this has occurred, make sure that the new layout has not caused design changes such as placing nozzle openings across, or too near to, welded seams. This is allowed in some codes, but is best avoided if possible.
- **Plate condition**. Check for dents and physical damage. Look for any deep grinding marks. Obvious grooves, deeper than 10% of plate thickness, caused by the edge of a grinding disc, are cause for concern.
- **Surface finish**. General millscale on the surface of the plate is acceptable before shotblasting. Check for any obvious surface rippling caused by errors during plate rolling.
- **Reduced thickness**. Excessive grinding is unacceptable as it reduces the effective wall thickness. Pay particular attention to the areas around the head-to-shell joint; this area is sometimes heavily ground to try to blend in a poorly aligned seam.
- **Bulging**. Check the whole shell for any bulging. This is mainly caused by 'forcing' the shell or head during tack welding to compensate for a poor head-to-shell fit, or excessive out-of-roundness of the shell.
- **Nozzle flange orientation**. Check that the nozzles have not pulled 'out of true' during fabrication or heat treatment of the vessel. This can cause the nozzle flanges to change their alignment relative to the axis of the vessel. A simple check with a steel tape measure is adequate.
- **Welding**. Make a visual examination of all exterior welding against the fabrication code requirements. You can get an indication as to whether the correct weld preparations were used by looking at the width of the weld caps. Check that a double-sided weld has not been replaced with a single-sided one, perhaps because the manufacturer has found access to the inside of the vessel more difficult than anticipated. Watch for rough welding around nozzles, particularly small ones of less than 50 mm in diameter. It can be difficult to get a good weld profile in these areas, so look for indications such as undercut, incomplete penetration or over-convex weld profile.

The vessel interior

It is important to make a thorough inspection of the inside of the vessel. This cannot be done properly by just looking through the manhole door; you have to climb inside with a good light to be able to make an effective inspection. Check the following points.

- **Head-to-shell alignment**. Most manufacturers take care to align carefully the inside edges of the head-to-shell circumferential joint, even though most codes allow other arrangements. Check what arrangement has been used and that there is a nice even weld-cap all the way around the seam.
- **Nozzle 'sets'**. Check the 'set-through' lengths of those nozzles protruding through into the vessel. Again, you should use the approved design drawing.
- **Weld seams**. Do the same type of visual inspection on the inside weld seams as you did on the outside. Make sure that any weld spatter has been removed from around the weld area.
- **Corrosion**. Check all inside surfaces for general corrosion. Light surface staining caused by the hydrostatic test water is not normally a cause for concern, but if the vessel specification does not call for internal shotblasting, such staining should be removed by wire-brushing. In general there should be no evidence of millscale on the inside surfaces; if there is, this suggests the plates have not been properly shotblasted before fabrication.
- **Internal fittings**. Check that these are all correct and match the drawing. In many vessels, internal fittings such as steam separators, feed baffles and surge plates are removable, with bolt threads being protected by blind nuts. The location of internal fittings is also important – make sure they are in the correct place with respect to the 'handing' of the vessel. It is also worth checking the fit of the manhole door and any inspection covers.

If you feel any uncertainty about the results of an internal inspection, it is best to address them immediately, before the manufacturer starts the process of preservation and packing of the vessel. Check any doubtful areas of welding for defects using a dye penetrant test. Small defects should be ground out. Make a note of all the areas you looked at, *describe* any indications and defects that you find, and make a location sketch.

The dimensional check

It is normal to carry out a dimensional check of pressure vessels during a final source inspection. Figures 8.8 and 8.9 show two typical examples. Although a vessel is not a precision item, the positions of the fittings attached to the vessel shell are important. Alignment of connecting

FIG 8.8
Checking vessel dimensions (1)

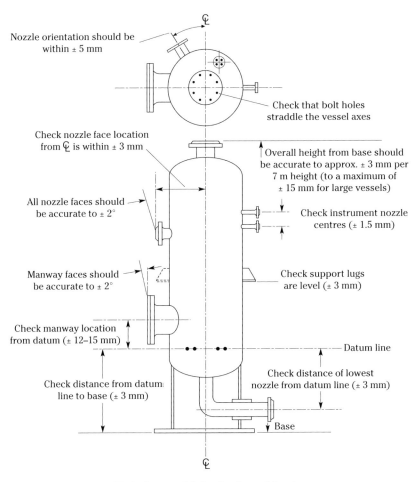

Typical acceptable levels of vessel 'bow'

Height	Diameter		
	<1200	<1300	>1700
<3000 mm	2.5	2	2
3000–9000 mm	4	7.5	6
>9000 mm	5	10	8

FIG 8.9
Checking vessel dimensions (2)

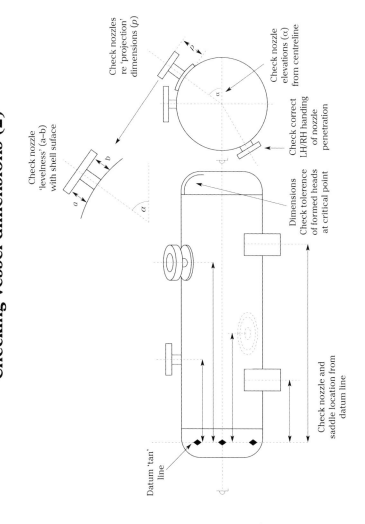

pipework can be affected and pipe stress calculations are done assuming correct alignment to fixed points such as vessel flanges. Misalignment must therefore be avoided. Practically, the dimensional check can be done either before or after the hydrostatic test. Any strains or distortions that do occur will be small, and difficult to detect by simple measurement methods. It is normal for the vessel manufacturer to have completed a dimensional examination report (this is a simplified sketch of the vessel showing only the key dimensions) before the SI arrives. This makes your task a little easier, eliminating the need to check against several different drawings. Dimensional checking can be done using a steel tape measure, with the use of a long steel straightedge and large inside or outside callipers for some dimensions.

Dimensional tolerances for pressure vessels tend to be quite wide. Follow the general tolerances shown on the manufacturing drawings. If any tolerances look particularly large, say more than +5–6 mm, double-check them; note that there is a technical standard DIN EN 13920 that gives general guidance for tolerances on fabricated equipment. Use the following guidelines when doing the dimensional check.

- **Datum lines**. First locate the datum lines from the drawings. Each vessel should have two: a longitudinal datum (normally the vessel centreline) and a transverse datum *(the tan line)*. The transverse datum is normally *not* the circumferential weld line – it is generally located 50–100 mm inwards from the seam towards the dished head (see Figure 8.9), and indicated on the vessel by three or four deep centre-punch marks.
- **Manway location**. Check the location of the manway with respect to the longitudinal datum line.
- **Manway flange face**. Check that this flange face is parallel to its indicated plane. A tolerance of +1% is acceptable.
- **Nozzle location**. An important set of dimensions is the location of the nozzles in relation to the datum lines. It is easier to measure from the datum line to the edge of each nozzle flange rather than to try to estimate the position of each nozzle centreline.
- **Nozzle flange faces**. Check these by laying a long straightedge on the flange face and then measuring the distance between each end of the straightedge and the vessel shell. It may also be possible to use a graduated spirit level in some cases. Nozzle flange faces should be accurate to within about 0.5% from their indicated plane. Check also the dimension from each nozzle flange face to the vessel centreline; a tolerance of +3 mm is acceptable.

- **Flange bolt holes**. Check the size and pitch circle diameter of bolt holes in the flanges. It is universal practice for bolt holes to straddle the horizontal and vertical centrelines, unless specifically stated otherwise on the drawing.
- **Vessel 'bow' measurements**. Both vertical and horizontal vessels sometimes *bow* about their axial centreline. This is the result of uneven stresses set up during fabrication and heat treatment. The amount of acceptable bow depends on the length (or height) and diameter of the vessel. A small vessel of 3–5 m long and a diameter of up to 1.5 m should have a bow of less than about 4 mm. A larger vessel of approximately 10 m long and 2.5 m in diameter could have a bow of perhaps 6–7 mm and still be acceptable. Larger amounts of bow than these approximate levels are generally undesirable. You can detect bow by sighting along the external surface of the vessel by eye. The extent can be measured using a taut wire; check at three or four positions around the circumference to obtain a full picture of the extent of any bowing.
- **Vessel supports**. Horizontal vessels usually have simple saddle-type supports. Check that these are accurately made so that the vessel sits level. A tolerance of $+3$ mm is good enough, perhaps a little less for longer storage vessels, which will have a greater tendency to bend if inaccurately mounted. Vertical vessels may have tripod-type saddle supports, or a tubular mounting plinth which fits over the lower end of the vessel shell (see Figures 8.8 and 8.9). Check this type carefully for accuracy – the vessel should stand vertically to within 1%. The best time to check a tubular plinth is before it is welded over the vessel shell.

It is not uncommon for a dimensional check on a pressure vessel to show a few dimensions that are marginally out-of-tolerance. These are not *necessarily* a case for rejection of the vessel. What is needed is a good sketch in your source inspection report indicating where the deviations are. You should say whether such deviations affect the function of the vessel (as would be the case, for instance, with excessive bow or major misalignment of nozzles) or whether they are merely cosmetic. Make specific mention of any dimensional inaccuracies that will have an effect on the amount of site work required to connect the vessel to its piping systems.

Vessel markings

The marking and nameplate details of a pressure vessel are an important source of information when a vessel is received at its construction site. A large plant can have several hundred vessels, constructed to different codes and for varying applications, so positive identification is a distinct advantage. Some pressure vessels will be subject to a further hydrostatic test after installation so information about the design pressure and works test pressure needs to be clearly shown. The correct marking of a vessel also has statutory implications; it is inherent in the requirements of vessel codes and most safety legislation that the safe conditions of use are clearly indicated on the vessel. Vessel marking is carried out either by hard-stamping the shell or by using a separate nameplate. It is preferable to use a separate nameplate on vessels with plate thickness of less than about 7 mm, or if the vessel is designed to operate at low temperatures. Figure 20.3 in Part B of this book shows the content and layout of an ASME VIII vessel nameplate. PED ones are different. Check the nameplate during the final inspection visit, using the following guidelines.

- The nameplate details should be completed *before* the vessel leaves the manufacturer's works. The only common exception to this is if it has been clearly agreed that a standard hydrostatic test will not be performed in the works but will be performed later at the construction site. This occasionally happens and, surprisingly, does not contradict the requirements of most vessel codes.
- If applicable, check that the statutory inspector has hard-stamped the nameplate. This will act as a general assurance to construction site staff that the statutory aspects of design and manufacture have been properly addressed.
- *Concessions*. A vessel manufactured with concessions from full code compliance must be identified as such. There are several ways to do this, the main criterion being that the concessions are brought to the full attention of the construction site and operation staff. Any resultant limitations on the use of the vessel can then be properly addressed. For some vessels, the normal way is to add the suffix *XX* to the vessel serial number on the nameplate, indicating that concessions are in force. If you see this designation, check that the technical details, and justification, of the concessions are adequately explained in the vessel's documentation package and make a special note in your inspection report.
- Make sure the nameplate is *firmly fixed* to the vessel. There should be a steel mounting plate welded to the shell and the nameplate bolted or riveted to it. Do not accept loose nameplates, as they will inevitably get lost.

It is the practice of some inspectors to take a 'pencil rubbing' of the completed vessel nameplate. This creates a permanent record which is included in the final documentation package. You can do this if you have nothing better to do.

8.7 Non-conformances and corrective actions

If non-conformances do exist in a pressure vessel, they are normally not too difficult to find and identify during a source inspection because the underlying requirement for statutory inspection encourages the use of detailed ITPs and a high level of inspection scrutiny. The problem is more one of corrective actions; non-conformances are often only discovered when the vessel manufacture is well advanced or complete, at which point the corrective action options can be limited. You will find two types of non-conformance: those where it is possible to initiate a corrective action to restore full specification or code compliance and those where the only action possible is a retrospective one. This second category rarely gives a perfect solution and may result in a permanent code concession being applied to the vessel.

Part of the role of an SI involved in pressure vessel inspections is to agree *solutions* to non-conformances. These take many forms but you can expect a lot of pressure to be placed on you by manufacturers to accept *retrospective* solutions, often involving some measure of compromise of the compliance criteria. For this reason you will find vessels a good 'proving ground' for the tactical approaches to source inspection mentioned in earlier chapters. Try to follow the principle of agreeing solutions to vessel non-conformances directly with the manufacturer; as long as you feel technically comfortable with what you are doing. This is the most effective and economical way to do it. The alternative approach, that of referring even minor points to technical specialists, may be acceptable, but the time and cost implications *will rise accordingly*.

Bearing these principles in mind we can look at 11 of the more common non-conformances that are found in pressure vessels. These are presented broadly in the order of frequency with which you should find them, starting with the most common.

Missing ITP documents

This is very common, caused by missing ITP steps, or the result of poor communication with sub-suppliers further down the manufacturing chain. You have to come to a quick decision as to whether a particular missing document really compromises code compliance or integrity. In about 90% of cases it probably will not – the *activity* will have been performed correctly but the document record will have been mislaid. The remaining 10% of cases will impinge on code compliance (these are often NDE-related) and lead to some uncertainty about weld integrity. The best steps to take are listed below.

- Decide whether code compliance or vessel integrity is in jeopardy.
- Ask questions. Look for evidence that indicates whether all the necessary activities have been done (in spite of the missing document).
- Make your best efforts to resolve the issues quickly. Contact sub-suppliers if necessary.
- Issue an NCR, making it clear exactly which document is missing. Give your decisions in your report, saying whether or not the missing document affects code compliance.

Incomplete statutory certification

There are numerous situations that can result in the conditions for statutory certification being incomplete. There may be outstanding design appraisal questions, missing documents, or observations made during the manufacturing process; all can cause the certifying organisation to be hesitant. For all SIs other than those representing the certifying organisation, this should be a straightforward issue. The objective is to obtain a certificate of *unqualified* code compliance, so place the onus on the certifying organisation to state clearly why it feels this is not possible. Ask for precise reasons, not just expressions of general discontent. Be careful to check that any vessel certificates that are subsequently issued are complete and unqualified – they should meet fully the wording of the applicable EN/ISO code or the ASME MDR (manufacturing data report) equivalent. It is best to be wary of certificates that have evasive wording or additional exclusions. If in doubt, look at the certificate under the assumption that a serious failure incident has occurred and you are required to *prove why* you accepted the certificate as an assurance that the vessel was fully compliant with the relevant code. We discussed at the beginning of this chapter that the vessel codes are stated in statutory documents as accepted examples of good practice, so it is poor practice to accept qualified vessel certificates.

Once again, the best solution is to find technical solutions quickly, before manufacture progresses to a point where the only solution involves a permanent non-compliance with the vessel code. Certification organisations are risk-averse and few will issue an unqualified vessel certificate on this basis.

Incomplete material traceability

Do not confuse incomplete material traceability with incorrect material *properties*, which is a different (more important) issue. Unless you have very firm evidence that code compliance *is* compromised (in which case a clear NCR should be issued), this is normally a retrospective exercise. For pressure shell components, the best step is to specify a re-test of the material specimens from the production test plate. This changes a documentation problem into an engineering activity. It allows a more objective solution than will an exercise of 'certificate-chasing' from material sub-suppliers, which takes a lot of time and will not re-instate any traceability chain that has been broken. Assess the re-test results carefully and explain in your report what was done and why. For non-pressurised components, such as vessel saddles and frames, you may wish to take a more relaxed view, to reflect the lesser effect these items have on the integrity of the vessel.

Incorrect dimensions

With large vessels it is not uncommon to find a few dimensions that are marginally outside the drawing tolerances. As long as any inaccuracy is not caused by serious distortions or bowing, the effects are generally of a minor nature. It is wise to exercise a little restraint – an NCR would only be properly justified if the vessel was clearly 'the wrong size'. For completeness, you should record the out-of-tolerance dimensions (using a sketch), particularly if these involve the nozzle positions, allowing any implications for the site connection of pipework to be assessed.

Head-to-shell misalignment

Maximum allowable head-to-shell misalignment is carefully calculated to keep the discontinuity stress (caused by the different response of the shell and head to internal pressure) within defined limits. These discontinuity stresses can become very high if the allowable misalignment is exceeded. If you find it, issue an NCR. There is little that can be done to rectify the situation short of remanufacturing the vessel. It is not advisable to pursue concessions to the vessel code in such instances – the technical risks are too high.

Incorrect weld preparations

The most common faults you will find are

- wrong weld preparation angles
- asymmetrical weld preparations machined the wrong way round
- incorrect root gaps (after tack welding).

It is difficult to make general statements about the acceptability of incorrect weld preparations. The weld preparation design is part of the weld procedure specification (WPS) for a particular welded joint, which is then qualified by the use of an approved procedure qualification record (PQR), showing the results of tests carried out on that configuration. Once this WPS/PQR link is broken by the use of an incorrect weld preparation, it is possible that the strength and integrity of the weld joint may be affected. As a general rule it is best to request that the preparation be re-machined to the correct configuration, if this can be done without removing so much of the parent material that it will cause mis-match of other joints in the vessel.

If re-machining is not possible, the best action is to specify that the incorrect joint be qualified (i.e. a new PQR), by making a test piece and subjecting it to the necessary non-destructive and destructive tests. This is particularly important for nozzle-to-shell welds as their strength characteristics are less predictable than for simple single or double-vee butt welds.

Incorrect weld procedure specifications (WPSs)

Treat this in a similar way to that for incorrect weld preparations, but anticipate more serious consequences. An incorrect WPS is often only discovered after the welding has taken place, so there may be no real chance to correct it without cutting the vessel and re-welding. The other solution is to try to qualify the actual WPS used by carrying out the PQR steps mentioned previously. Do not be too enthusiastic about recommending this action, however – take a close look at the incorrect WPS first. If the modifications include any major changes to the weld root (for instance the lack of a GMAW/MIG root-run where this has been specified previously), or to the filler (consumable) material, it is unlikely that any subsequent attempt to qualify the new WPS will be successful. It will probably fail the tests. For other errant weld variables, the technical risks involved are less significant in terms of consequences. Make it clear in your report that you have taken the essential variables into consideration before making your decision.

Incorrect material properties

This refers to a situation where you find the material properties are *marginally* outside the specification tolerances, not a situation where a totally incorrect material has been used in error. On balance, the mechanical properties of a pressure vessel material are more important than its chemical composition. This is because the mechanical properties are a *function of* the chemical analysis and because mechanical properties can be changed by heat treatment, which leaves the chemical analysis nominally unchanged. Here are two main types of non-conforming mechanical properties that you may encounter.

- **Out-of-specification tensile properties**
  The most common problem is that the yield strength is too low. For most pressure vessel designs, marginal differences in yield or tensile strength can frequently be compensated for by the factor of safety already built in to the design. A corrosion allowance will also be included, although under most code rules this must not be included in the material thickness value used for stress calculations. Marginal differences in tensile properties alone should rarely cause a vessel to be declared unfit-for-purpose. In the first instance your NCR should specify a recalculation exercise to demonstrate whether the revised factor of safety (using the actual material yield or tensile strength) meets the code requirements. Assess the results carefully in conjunction with the certification organisation.
- **Out-of-specification impact or hardness properties**
  For these, the effects are likely to be more serious. Low impact values or high hardness readings imply increased brittleness of the material. Brittleness is related to crack propagation, one of the mechanisms that contribute significantly to practical failure mechanisms. A sound strategy is to specify a programme of re-tests, even for marginal out-of-specification results. Remember that a minimum of three specimens is required, because of the inherent difficulties in obtaining reproducible results from impact tests (see Chapter 5). If the re-test results confirm that the impact properties are genuinely too low, this casts a serious shadow of doubt on the integrity of the vessel. It is not easy to demonstrate *explicitly* the effects of poor impact strength, so it is difficult for vessel designers to demonstrate acceptability of out-of-code results in the way that is possible with reduced tensile properties.

You can minimise wasted time and effort by anticipating these outcomes in advance. As an SI you will be well justified to reject a vessel that has out-of-specification impact properties. If you can demonstrate firm evidence of increased brittleness, you are on safe technical ground and it is unlikely that the certifying organisation will overrule this type of decision. Be wary of proposed solutions that advocate derating the vessel to operate at a lower pressure. There is little technical coherence in such arguments if low impact strength is the problem.

Missing NDE

Missing NDE is an important omission, but not one which should cause too many problems. The solution is simply to specify that the NDE be repeated on the vessel in its current condition. Surface crack detection using PT or MT is straightforward and UT can replace RT under most vessel codes. Specify the technique to be used, and the relevant acceptance standard, in your NCR. There are few valid reasons why UT cannot be done on a completed pressure vessel; any rough surface finish can be improved by grinding and the critical areas of butt and nozzle welds are easily accessible by a skilled operator (see Chapter 7). It is poor practice not to ensure that all the specified NDE tests are properly carried out, as NDE results are important evidence that the integrity of welded components has been properly assessed. Their review is an important SI responsibility.

Remember the common misconception (introduced in several areas of this book) that a hydrostatic test can be considered as a substitute for missing NDE. It is worth repeating that, although the 'surface arguments' for this approach can seem convincing, the underlying technical rationale is poor. Hydrostatic tests may identify leakage, but they will not necessarily reveal the type of defects that contribute to the onset of failure. The correct code-specified NDE is essential to demonstrate the integrity of the vessel.

Unrecorded repairs

Occasionally during the visual examination of a vessel you may find evidence of unrecorded repairs. This is more common in welds than in the body of components such as nozzles. Repairs are allowed under all manufacturing codes, but should have been recorded, along with details of the repair procedure used. First, check whether the repair procedure and records really are unavailable – they may exist, but just not be included in the document package submitted for review. If you conclude that a repair is genuinely unrecorded then you are justified in taking

FIG 8.10
Watch out for: Pressure vessel inspections

Fitness for purpose

1. The overriding criteria are the *code compliance and integrity* of the vessel – particularly the welded joints. The industry has developed four commonly accepted activities to try to provide an assurance of integrity. They are
   - an independent design appraisal
   - material traceability
   - prescribed NDE activities
   - hydrostatic (pressure) testing.

Certification

2. The requirement for vessel certification may be imposed by an end-user, purchaser insurance company or statutory authority.
3. Vessel certification *only* shows that a vessel complies with the design code (EN, ASME, TRD etc.). It is not a guarantee of integrity or fitness for purpose. It is normally unrelated to project-specific requirements.

Pressure vessel codes

4. Vessel codes deal more with design than source inspection. Much of the inspection-related information you will need is contained in referenced documents, not in the code document itself.
5. Expect to see vessel codes used for other types of fabricated and cast components. In such cases a more flexible approach is required.
6. A hydraulic (hydrostatic) test is a test for leakage, and a vessel's ability to withstand static principal stresses. When vessels do fail, they normally do so because of *different* failure mechanisms, so a hydrostatic test is not a full substitute for the correct NDE.

Final inspection

7. The final visual/dimensional examination of a vessel benefits from a careful, structured approach. Use a checklist.
8. There are several well-known categories of non-conformance that you should be aware of. Some have only retrospective (and imperfect) solutions if discovered too late in the manufacturing process. If you are not careful (or are badly organised), the *costs* of deciding corrective actions will rise steeply.

further action to make sure that no *other* similar repairs exist, and to check the integrity of those unrecorded repairs that have been done. Start with 100% surface NDE, followed by a percentage volumetric examination in the most critical areas (butt weld tee-joints and those full-penetration weld types shown in Chapter 6 of this book). Make sure that your NCR specifies clearly the type and extent of NDE you feel is necessary.

Hydrostatic test leaks

If there is any leakage at all during a test on vessels, pipeworks or valve bodies, even from temporary flanges or pipe connections, issue an NCR. Leakage from welds is usually an indication of either extensive porosity or cracks. The question you must ask is: *why were these not discovered during the NDE activities*? You need to decide whether there are other unseen risks to the integrity of the vessel, in addition to those defects that are causing visible leaks. Areas of porosity and cracking can be excavated and then re-welded. Make sure that the repair procedures are properly reviewed and approved first, as some materials will require further heat treatment of the whole vessel to eliminate stresses induced during the repair welding. Record all repairs using sketches if necessary.

Chapter 9

Inspecting valves

Pipes and valves are used in all fluid systems – even a small process plant will contain a large variety of types. For large offshore and chemical plant projects these two sets of components can often account for 10–12% of the overall contract value. They are best considered separately for inspection purposes. Valve testing is an important, if rather routine, part of works inspection. There are numerous different types of valves; however, the principles of inspection are much the same for all types. Many valves have cast bodies, often with weld-prepared ends so the material principles discussed in Chapter 5 are of direct relevance. Valves are also subject to volumetric and surface NDT techniques: RT, UT, PT and MT are all used, depending on the specific application. The principles and techniques are the same as those discussed in Chapter 7.

Valves are well covered by technical standards. In general, the standards are concise and easy to use, although, like piping, they can contain a lot of design-related information that is not directly relevant to source inspection activities. Two of the most common standards are ASME B16.34 and ANSI/FCI 70-2.

ANSI/FCI 70-2 is an American National Standard which specifically covers seat leakage of control valves. It specifies six classes of allowable seat leakage ranging from Class I (zero leakage) to Class VI, in which a significant amount of leakage is allowed. You should see this standard used quite regularly – so make sure you know what the different leakage classes mean.

API 598 is a standard for body and leak testing of valves. This is covered in Chapter 21 in Part B of this book.

9.1 ASME B16.34

ASME B16.34 is a popular valve standard used in many countries. It is used in conjunction with the ASME codes on boilers and power piping in ASME design installations, but also quite separately as a valve standard in itself. You will often see it used as a reference standard on plant built to non-ASME guidelines. The standard is relevant to flanged, weld-neck and threaded end types for all applications – it is not industry specific. Theoretically, it covers all valves down to the very smallest sizes but, in practice, its main application is for those above about 60 mm nominal bore.

The B16.34 document is quite thick and, in common with other US-style standards, provides wide technical coverage that includes design aspects. The parts relevant to source inspection are spread throughout various sections of the document and the main ones that you need to know about are

- the different *valve classes*
- materials and dimensions
- NDT scope
- NDT techniques and acceptance criteria
- defect repairs
- pressure testing.

Valve classes

ASME B16.34 divides valves into two main classes: *standard* class and *special* class. There is a third, historical one called *limited* class, but it is not used very often. There is then a series of pressure–temperature ratings for each type, designated as class 150, 300, 400, 600, 900, 1500, 2500 and 4500, which are related predominantly to the valve inside diameter and its *minimum* wall thickness. The higher the number is, the larger is the wall thickness and the maximum design pressure.

The class numbers are all listed in a table of the standard and rarely cause any controversy during inspection; it is merely a matter of checking the measured wall thickness against the table. Most of the problems arise from the broader issue of whether a valve is defined as standard or special class. The main difference between the two is in the amount of NDT required, as special class valves have to pass a set of NDE tests that are well defined in the standard. These are fairly onerous requirements which many mass-produced cast valves have trouble in meeting. In contrast, for valves designed as standard class, ASME B16.34 does not specify any volumetric NDE, leaving the verification of the integrity of the valve to the pressure test and/or any agreement reached between the purchaser and the manufacturer.

The difficulty arises when you look for a definition regarding which types of valves should be designated as special class. ASME B16.34 does not tell you this – it does not, for instance, state that valves installed in gas hazardous chemicals or similar safety-critical systems need to be special class. The result is that the purchase specification or contract is the only place you can look for guidance on this point. If a valve is not overtly specified as special class, you cannot insist that it needs all the NDE tests in B16.34 (even if you think it *is* a safety-critical application). This situation is not uncommon and has caused integrity problems in some plants. One specific restriction given in B16.34 is given below.

- Flanged end valves (where the flange is cast integral with the body) cannot be rated as 'special class'.

Materials and dimensions

The standard does not contain any information against which it is possible to check the material selection that has been used for a valve. For this, you need to look at one of the referenced ASME standards, such as ASME B31.3. There is a materials table in B16.34, but it is simply a summary of the ASME materials with their product forms and P-numbers. There is more information provided about valve *dimensions*. Main areas covered are listed and described below.

- *Wall thickness*. The important criterion is the minimum wall thickness of the valve body, excluding any liners or cartridges that may be fitted. There is useful general information on exactly where on the valve body this minimum wall thickness is measured, although you will need to interpret it in the context of the particular valve design being inspected.
- *Inside diameter*. This again has to be measured to see if a valve complies with its class rating, that is, 150, 300 etc. A section of the standard explains how this is determined.
- *Neck diameter*. Valve body necks also have to meet the minimum wall thickness requirement shown under a valve's class number. There are two slightly different ways to calculate this, depending on whether the valve is rated as 2500 and below, or above 2500. This can be a tricky task (as there are so many different designs). Practically, it is normally sufficient to concentrate on the wall thickness, unless you have serious concerns about the geometrical arrangement around the neck area.

Scope of NDE

NDE is often the most relevant part of B16.34 from a source inspection viewpoint. Both volumetric and surface NDT scopes are specified (for

FIG 9.1
NDE of ASME B16.34 valves

Gate valve

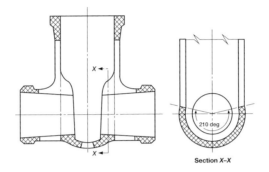

Areas for volumetric NDE

Butterfly valve

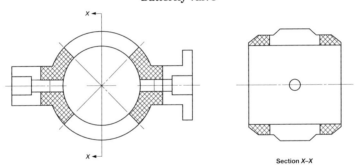

Globe valve

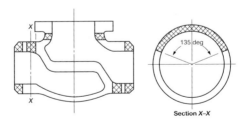

Illustrations acknowledgement: ASME B16.34

special class valves only, remember). As with some pressure vessel standards, both radiography and ultrasonic techniques are acceptable. The essence of the NDE scope is that it concentrates on those areas of the valve in which defects are either most likely to occur, or would have the most detrimental effects. For *volumetric* NDE, the areas to be tested are shown in Figure 9.1. They concentrate on the weld end, bonnet, neck and the junction between the body shell and the seat. The exact location and size of the areas to be tested differ between designs, and the standard shows the application for several examples. In practice, the differences are slight, so you can safely use Figure 9.1 as an approximate model to follow. The scope is the same whether RT or UT is used.

Specification of surface NDE of special class valves is simpler; B16.34 requires that all external and accessible internal surfaces of the body and bonnet castings be checked for surface defects. Materials classed under ASME Group 1 can be tested using MT and PT, whereas non-magnetic Group 2 and 3 materials must be tested by PT.

Techniques and defect acceptance criteria

RT procedures for cast valves are the same as those specified for normal ASME applications, for example, ASTM E94 and ASTM E142. There is an additional requirement that every radiograph has a penetrameter and the defect acceptance criteria to be used depend on the wall thickness.

Ultrasonic techniques are the same as those used under the general standard for UT of castings: ASTM A609; however, B16.34 does quote its own defect acceptance criteria.

Note that these defects can become fairly large before they become unacceptable. Many experienced purchasers of valves for safety-critical applications impose their own more stringent defect acceptance requirements. This is good practice.

Surface examination techniques are the same as for castings: ASTM E709 for MT and ASTM E165 for RT. Both have the same acceptance criteria for linear and rounded indications.

Defect repairs

This can be an awkward issue from a source inspection viewpoint. In common with most other published standards on castings, B16.34 places no limit on the number and extent of defects that can be repaired by welding. The only constraints are those related to the welding technique itself; that is

- WPS and PQR to be in accordance with ASME IX
- heat treatment to comply with ASME VIII (Div 1)
- repaired areas need to be re-examined using the same NDE technique
- defect acceptance criteria for radiographed repairs to be in line with ASME VIII Div 1 UW-51; the same as used for pressure vessels.

See Chapter 21 in Part B of this book for more information on valve testing.

Chapter 10

Inspecting painting

You will find that almost everyone likes to comment on painting. It is visible, its purpose is well known and so it always attracts attention during source inspections. Despite this attention (a subtle mixture of the authoritative and the uninformed) you will still hear reports from site engineers and end users about poor painting. Often it is only a cosmetic problem, but sometimes it is more serious and the paint flakes off after a very short time, allowing corrosion to start. One reason for this is poor source inspection.

For small, robust items of fixed equipment, such as valves, fittings and small vessels, the main reason for painting the item is largely cosmetic rather than protective. For large and fabricated items operating in a corrosive environment, however, the priorities change. Structures for offshore, coastal and similar use very specialised paint systems to reduce the effects of corrosion. Painting has become almost an integral part of these technologies. The main specification integrity/acceptance criteria are

- proper preparation
- a suitable paint system
- correct application.

Note how these are composite criteria; they refer to the *steps* of the painting process rather than purely to the assessment of the integrity of the paint after it is applied. This is similar to the approach used for the assessment of welding in Chapter 6 of this book. Put simply, effective source inspection of painting is about monitoring the painting *process*, not just the end result.

To be able to inspect paintwork effectively it is necessary to understand the role of the paint coating. This role differs, to a limited extent, between types of paint, related to the chemistry of the paint system. This is covered in detail in many standard text books so is not repeated here. What you *do* need to know as a source inspector (SI) is the

FIG 10.1
Practical reasons for paint failures

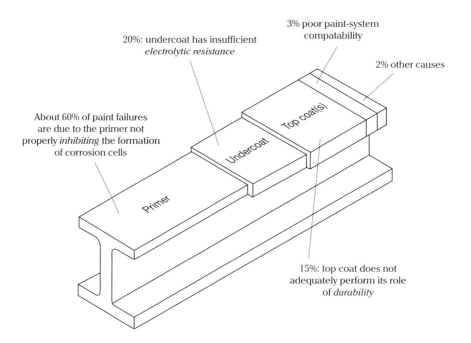

likelihood of finding problems that will result in failure of the paint film. Figure 10.1 shows the approximate frequency of the various painting problems that you can expect to find over a large number of source inspections of fixed equipment. Note how the problems are shown in terms of the role of the paint that is not being fulfilled in each case.

10.1 Paint types

There are numerous different types of paint. For the purposes of fixed equipment source inspection, paint type is only important insofar as it influences the methodology of the inspection, and the way in which it affects the specification acceptance criteria. This is a simplified but satisfactory approach; there is no need to have detailed knowledge of the chemistry of the subject. It is useful, however, to try to learn a little about the major classifications to which paints belong, rather than to rely solely on manufacturers' product trade names, which can be confusing. The main classifications are outlined below, along with the main features that have an effect on their inspection.

Air-drying paints

Air-drying paints are the most common type that you will meet in fixed equipment applications. They account for maybe 70–80% of all painting on indoor structures and equipment, or on outdoor installations which are not subject to highly corrosive atmospheres. There are four main types.

- **Alkyd resins** are used in primers and undercoats. Dry film thicknesses (dft) are generally 35–50 µm per coat. Expect to see cosmetic problems caused by the short-term *wet edge time*. Adhesion is not a common problem if preparation is done properly.
- **Epoxy esters**. Expect to find these on structural steelwork and storage tanks. They have better chemical resistance than all alkyd types. The inspection requirements are the same; however, you may need to guard against problems caused by poor spraying technique with these type of paints.
- **Chlorinated rubbers**. These are used for exterior protection of structural steelwork and fabrications in particularly corrosive environments such as coastal or offshore locations. The top coat is normally applied over a similar chlorinated rubber alkyd or zinc-rich primer; other types may result in adhesion problems. Coat thicknesses vary, but single coat dry film thicknesses are normally quite thin, about 50–60 µm per coat owing to the high level of solvent in the paint. Often, however, specifications will call for a single thick coat of 300–400 µm. For this, a thickening agent has to be added to the paint. Chlorinated rubber paints do not often suffer from inter-coat adhesion problems, as subsequent coats fuse together quite effectively. They *do* suffer from problems such as sagging and peeling if the coating is applied too thickly.

Two-pack paints

Two-pack paints comprise a pigmented resin and a catalyst or hardener, which are mixed together and harden via a chemical reaction rather than evaporation of the solvent. The most common range is the epoxy type, which is highly resistant to water, environmental and chemical attack. Expect to see it specified anywhere where there are alkali or acidic conditions rather than for general service use. Note the following four specific source inspection points.

- Epoxy paints are generally 'hi-build'; an average dft is 100 μm per coat.
- They have a short pot life once mixed, so application problems do occur.
- The ambient temperature during painting is important. Epoxy will only cure successfully at temperatures above 7°C.
- On balance there are more critical process-related factors than for air-drying paints. This means that monitoring of the application process is an important part of the source inspection process.

The other type of paint in this category is two-pack polyurethane, generally used for chemical-resistant applications. Dry film thickness can vary from 40 to 100 μm per coat, depending on the paint formulation used.

Primers

For metal which has been shotblasted it is common for a thin (30 μm) coat of zinc-based blast primer to be applied. This can be an 'etch' type containing phosphoric acid to etch the surface, or a two-pack epoxy rich in metallic zinc. Zinc primers have inhibitive properties, the zinc providing cathodic protection to the iron. From a source inspection viewpoint a key issue is to ensure that these primers are applied immediately after shotblasting. The zinc must make intimate contact with the metal surface and not be restricted by corrosive products, which can form very quickly on a freshly blasted surface. For less critical fabrications, you may find zinc chromate or zinc phosphate primers being used. These are traditionally applied in quite thick coats and tend to be more common for on-site repair work rather than for new factory-built equipment. Their performance on steel that has not been fully cleaned is quite poor.

Preparation

Incorrect surface preparation is the most common root cause of failure of paint films, particularly those applied over common ferrous materials such as low-carbon steel. Poor preparation does not always result in instant failure; a period of 2 or 3 years may elapse before the real problems become apparent. By then, however, the breakdown will likely be almost complete, the paint system having effectively given up its protection of the underlying metal.

Proper preparation comprises two objectives. Removal of existing rust cells is one, but the main one is to eliminate any active (or latent) corrosion cells on the surface of the metal by removing the *millscale*. Millscale is a hard, brittle oxide formed during the steel rolling process. It causes mechanical problems with a paint film because it expands and cracks, causing the paint to flake off. It also causes electrochemical problems because it is cathodic to steel, so it encourages rapid anodic attack on small areas of unprotected surface. Note the following three key points about millscale.

- Proper preparations means elimination of *all* the millscale before painting.
- Millscale cannot be removed by wire brushing, it is too hard. The practice of 'weathering' steel by leaving it outside only reduces the amount of millscale, it does not remove it totally.
- The only practical way to remove millscale properly is by shotblasting with the correct heavy grade of grit.

The degree to which a material is shotblasted is important. Grades of cleanliness of the blasted surface are covered by several well-accepted standards (see Chapter 23 in Part B of this book). From a SI viewpoint, however, it is wise to be wary of grades of finish that do not specify complete removal of the millscale. See Figure 10.2, which shows some key source inspection points that you should check.

Application

Most fabrications, vessels and larger equipment items that you meet during source inspections will have their paint applied by spray. Methods such as dipping (for small components), hand brush or roller application, and electro-deposition are less common. The technique of spray painting is one which relies heavily on the skill of the operator; it is perfectly feasible for a properly prepared surface, with a well-chosen paint system, to give poor results because of a poor application technique. This is a specialist area; however, as an SI it is useful to have an appreciation of the main variables that have an influence on application. You will find them addressed in the painting procedures and record sheets used by painting contractors. Use them also as a rough checklist of items to look at when you are *witnessing* paint application. The main variables are listed below.

FIG 10.2
Shotblasting – Points to check

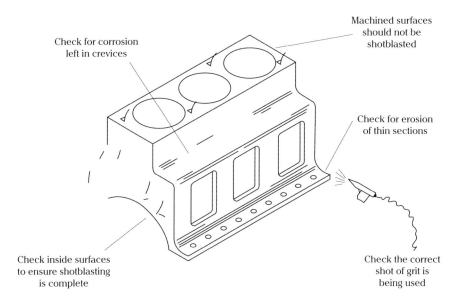

Shotblasting: Points to check

- Ideally, all millscale should be removed
- Check for areas of over-blasting, where critical dimensions have been eroded
- Make sure the blasted surfaces are brushed or vacuumed before priming
- Primer should be applied immediately after blasting. A quick-drying 'blast primer' should be used if there is any delay before the main primer is applied

Preparation grades (EN codes)

Sa 3	Blast cleaning to pure metal. No surface staining at all
Sa 2½	Very thorough blast cleaning. Only slight surface staining may remain
Sa 2	Thorough blast cleaning. Most millscale and rust is removed then the surface cleaned
Sa 1	Light blast cleaning. Any loose millscale and rust is removed

Other standards that may be used are SSPC (see Part B of this book)

- **Spraying air quality**. It should be moisture free, by using filter/dryers, in order to avoid contamination of the paint. Some techniques use an airless spray in which the paint atomises due to pressure drop only as it exits the spray gun.
- **Paint mix and consistency**. This is particularly important for two-pack paint types. Shelf life and pot life restrictions need to be complied with. Note that some paints are heated before being applied.
- **Ambient conditions**. This is a key variable. Ambient temperature and relative humidity must be within prescribed limits (paint specifications clearly state what these are). High humidity (above 80%) and low temperatures (below about 4°C) are cause for concern. Dust-laden or salty conditions are also undesirable.

10.2 Paint specifications and standards

The extent to which technical standards are quoted in contract specifications depends on the purpose of a piece of equipment, and the practices of the industry within which it is used. For petrochemical plant and general industrial uses it is unlikely that you will need to become familiar with the detail of many of the specific technical standards; source inspection of painting is more about careful observation and the application of a certain amount of judgement. In short, a common-sense approach works best. The corollary of this, however, is that you cannot rely heavily on technical standards to reinforce your judgements. Experience will help; this is another reason why it is important to witness painting techniques at the preparation and application stages. It is much easier to understand the techniques of application once you have seen them being done.

There are several sets of standards relevant to surface preparation and paint testing. The most common ones are given below.

ISO 8501/BS 7079 covers the generic subject of preparation of steel substrates before painting. It comprises 16 separate documents divided into four groups, as follows.

- Group A covers visual assessment of surface cleanliness.
- Group B covers more general assessment of surface cleanliness using chemical and pressure-sensitive tape methods. It is unlikely you will see these used during normal works inspections.
- Group C covers assessment of surface roughness (or profile) using comparative and direct instrument measurement methods.

- Group D covers the preparation techniques themselves, including shotblasting and power brushing.

Swedish standard SIS 055900 (ISO 8501) is a similar standard specifying 'grades' of shotblasted surface finish. It provides reference photographs for four main grades of blasted finish, termed Sa3, Sa$2\frac{1}{2}$, Sa2 and Sa1. You will frequently see these quoted by themselves, without explicit reference to the SIS 055900. Figure 10.2 shows the interpretation of these grades. Nearly all contract painting specifications require that surfaces are prepared to at least Sa$2\frac{1}{2}$. To meet this grade the surface must be blasted to the state where only *traces* of millscale or rust remain. In practice this is normally interpreted as meaning that only a light staining can be seen – it is only slightly short of the Sa3 level, which requires that the surface exhibit a pure white metal, totally uniform finish.

SIS 055900 also makes provision for describing the degree of rusting of hot rolled steel before it is shotblasted. There are four grades A to D. Grade A is where the steel surface is still fully protected by millscale with little, if any, rusting, whereas grade D represents the worst condition in which the millscale has rusted away and the exposed surface has become heavily pitted. You may see these grades shown in conjunction with the blasted finish designation – for example, C Sa$2\frac{1}{2}$ represents a moderately rusted surface which has then been shotblasted to Sa$2\frac{1}{2}$.

There is a similar American standard from ASTM SSPC (Society for Protective Coatings). This uses similar principles of describing rusting and blasted finishes but uses different designations. The most commonly used ones, A, B, C or D Sa$2\frac{1}{2}$, are designated as grade SSPC-SP10. This is referenced in Part B of this book as it forms part of the API source inspector of fixed equipment (SIFE) examination body of knowledge.

BS EN 12944 *Corrosion protection of steel structures by paint systems* is a very broad technical standard of eight separate parts. It covers painting requirements for most of the engineering applications you are likely to meet. There is clear guidance on how to choose and specify a paint system suitable for various environments – it provides some data on the performance of various types of paints but does not go far enough to enable a full assessment to be made from the information provided.

The issue of environmental conditions is covered in BS EN 12944: Part 2. Various environments are defined, ranging from clean indoor conditions to a polluted coastal atmosphere. You will see these commonly referred to in the parts of contract specifications dealing with the choice of paint systems.

The issue of paint colour is dealt with by a series of standard colours known as 'RAL numbers', originally from a German standard. These RAL samples have a four-digit designation and a simple swatch card is available for colour comparison purposes. Most of the colours are not too hard to differentiate from each other; however, you may have a little difficulty with the blues and greens (RAL 5000 to RAL 6000), which use a number of quite similar shades. By all means use a swatch card during your works inspections – but be prepared to allow a little leeway regarding the exact shade of paint that is used.

The determination of dry film thickness is covered by ISO 2808 and by ASTM D1186 (see also SSPC standards covered in Part B of this book).

10.3 Test procedures and techniques

There are only a small number of test techniques that are practically used during source inspection of paintwork. They are only partially useful in verifying the specification compliance and integrity of a paint film – there are some very real paintwork problems that they will *not* detect. This means that effective source inspection of paintwork consists of verifying the three main criteria ('preparation', 'system' and 'application', as introduced earlier in this chapter). After this it is heavily dependent on good *visual* examination, assisted by simple tests.

Visual examination

A thorough approach is required to the visual examination – the objective is to find any major problems that exist. Before you start, make a point of visualising how the operator would have sprayed the component. It should be reasonably clear which way up the item would stand, and whether the operator would spray by standing at ground level, or whether scaffolding or a mobile gantry crane would be needed (for large fabrications and structures). This will provide insight into where potential problem areas may lie. Check the following areas. They are equally valid at the primer, undercoat or top coat stage.

Enclosed areas
Check these first. Enclosed areas are those that, although not necessarily totally enclosed, cannot be sprayed directly from 'outside'. Typical examples are the inside of tanks or sumps, box section girders, hollow

sections with a lot of internal ribs or stringers, and ductwork; all of these require an operator to reach *inside* to apply the spray. Visibility and access will be difficult, so such areas are often the easiest place to find problems with both surface preparation and paintwork application.

Inside corners and radii
Inside radii often have problems with excessive paint application, causing it to sag and peel off. Check also the inside corners of fillet welds, particularly enclosed corners where plate material joins together in three planes (as in the inside corners of a box). Where such areas are some distance from the operator's spray-gun, because of difficult access, look for the opposite problem – these areas are often too thinly coated because the operator is deliberately trying to avoid applying too much paint.

Inside edges
These are edges that face *away* from the operator during spraying, normally because there is no access to the other side. They are not easy areas to spray. Look for too-thin dft on the edges.

Outside edges
Outside edges are those that face *towards* the operator and as a result are more predictable. A good spraying operator will coat the outside edges first, before covering the horizontal or vertical areas behind them, so look for areas of incomplete coverage on the edges, to see whether this has been done properly. It is fairly common to see paint ridges perhaps 100–200 mm back from the edge (check by sighting along the edge). This is caused by the operator allowing the wet edge to dry slightly before re-covering it when spraying the nearby areas, or by the application of a 'stripe coat' to build up the edge thickness. There is no need to be too concerned about this; it is only a threat to coating integrity if the edge dft is so excessive that recognised over-thickness defects such as paint sagging or runs start to appear.

Large horizontal surfaces
Check that the coating looks even without any excessively thick or thin areas. Problems can sometimes result from an operator trying to spray from one level, without the correct scaffolding or gantry. Large horizontal planes are a good place to find defects caused by poor ambient conditions. Look for blistering or uneven thickness caused by condensation.

Large vertical surfaces
Large vertical surfaces are normally easier to access for spraying, so unevenness is not a common occurrence. The main problem is paint *runs*; vertical surfaces are a good test of the consistency of the paint mix (particularly with two-pack types) and of the skill of the operator in avoiding over-application.

Around fittings
Most fabrications have integral fittings such as saddles, nozzle stubs, reinforcing rings, hatches and lifting/support fixtures. Check around them carefully as some have reverse edges, webs and fillets which can be overlooked by the operator when spraying. You are more likely to find paint defects due to the paint film being too thick in these areas, rather than too thin.

Masked-off areas
It is common for various areas such as flange faces and precision machined mating faces to be masked off during painting. It is worth checking the manufacturing drawing to make sure that the marked areas are indeed the correct ones. Equally, be very wary of any milled, turned or ground surface which has been painted over – this is *almost certainly* a mistake.

Painting defects

The main painting defects that you are likely to see are described in the following subsections.

Sagging and curtaining
These are easy to identify, consisting of obvious areas of paint over-thickness, where the wet film has sagged under its own weight. You will most often see it in wide horizontal bands on large vertical surfaces. It is caused by applying too much paint. In extreme cases it will develop into *runs* down vertical surfaces. Although unsightly, it is not necessarily cause for rejection, unless it is widespread or is accompanied by surface preparation problems.

Orange peel effect
In this case, the paint surface when dry resembles orange peel, with a dappled surface, sometimes accompanied by small blisters. This may be a problem of paint consistency or poor spraying technique. In most cases the adhesion of the film is affected, so significant areas of orange peel effect are a clear cause for rejection.

Wrinkling or lifting
At first glance this appears similar to the orange peel effect, with a blistered appearance to the film surface. The quickest way to differentiate is to run your hand over the surface to check whether any significant lack of adhesion is present. If the surface film is clearly loose (especially at the edges of any large blisters) then this is indicative of an *intercoat* problem. The most serious type is caused by incompatibility of the solvents used in successive paint coats, in other words, a paint system error. A similar, but less pronounced, effect can be caused by the operator not allowing the correct time interval between coats.

Rough surface finish
A poor surface finish is normally indicated by the loss of gloss on the top coat and is easiest to see on large, flat surfaces. The main causes are condensation or airborne dust and dirt. Unless dirt contamination is very serious, it should be possible to rub down and re-coat, without having to remove the existing paint coats.

Pinholing
This has the appearance of large numbers of small, concentrated pinholes in the paint surface. The most likely cause is contamination of the paint by oil or water. Extensive pinholing does require a full repaint, otherwise the fitness for purpose of the paint system is likely to be affected.

Thin areas
Too-thin coating is the most common defect that you will encounter; expect to find components with overall too-thin coatings as well as those where only individual areas such as edges have insufficient application. Provided that the primer and undercoat films have been applied, then the top coat thickness can easily be built up by further application of paint. If, however (as occasionally happens), the primer or undercoat are much too thin, or missing, then the coating will not meet its specification requirements. Strictly, an additional thickness of top coat is not an acceptable remedy for a deficient primer layer; primer often has an *inhibitive* role whereas the main function of the top coats is to protect against weathering and mechanical damage.

Dry film thickness (dft)

The dft of a paint coating is an important parameter. Assuming correct preparation and application, the durability and protective properties of a coating are related directly to its dft. Adequate thickness is necessary in order for the film to have sufficient electrolytic resistance to prevent the formation of the local galvanic cells that cause corrosion. Most painting specifications show the dft required for the separate primer, undercoat and top coat, as well as for the three together. You will see that these are shown as *minimum* dft. Sometimes a maximum is also quoted, but often it is not; it is inferred that a thicker dft is acceptable, as long as this is not so thick as to result in peeling, sagging or other defects associated with excessive application. It is only possible to obtain a true dft reading when a paint film is completely dry; this can be up to 24 h after application of the last coat, depending on the type of paint. During application, the painting operator checks wet-film thickness using a comb or wheel gauge. This thickness then reduces to the dft as the solvent evaporates. It is difficult to quote general relationships between wet and dry film thickness as this varies with the type of paint (it will be shown on the paint manufacturer's data sheet if you need to know it when you are witnessing paint being applied).

A simple hand-held meter is used to measure dft. These work on electromagnetic or eddy current principles and provide a direct digital readout. Some have a recording and print-out facility. There are a few general guidelines to follow when witnessing dft measurements, as given below.

- **Calibration**. Do a quick calibration check on the dft meter before use. Small test pieces, incorporating two or three film thicknesses, are normally kept with the meter. You can also do a quick zero calibration test on a convenient exposed steel surface; a machined surface is best.
- **Test areas**. Test a large number of points, as two or three are not sufficient. Make sure you include vertical and horizontal areas, corners, radii and edges. Pay particular attention to enclosed areas. Use your appreciation of which areas would be difficult for the operator to spray easily.
- **Results**. Dft requirements specified are generally accepted as being considered nominal or average values. As a broad industry rule, the average dft measured (from multiple readings) over a minimum 1 m^2 area should be greater than or equal to the specified level, *but* there should be no individual dft readings of less than 75% of the specified level. This means that there is always room for some interpretation, unless a purchasing specification is very specific about the number

and locations of measurements that are to be taken. See Chapter 23 of this book for details of an SSPC standard covering this.

If in doubt, the best standard to look at is ISO 2808 or ASTM D1186.

Paintwork repairs

The easiest place to do repairs to paintwork is in the works, before shipping the equipment to the construction site. Some construction sites are well equipped and have the necessary skilled subcontractors for painting large items, but many are not, and construction sites have environmental problems such as high humidity, dusty or salty air. Repair techniques depend on whether the faults you have found constitute a major specification or integrity non-conformance (wrong preparation, system or application remember) or whether they are cosmetic. A situation involving a major non-conformance normally has only one solution; remove the faulty paintwork and repeat the process, taking care to eliminate the previous faults. Given that serious faults have *already* occurred, there are several points that you should consider.

- Do not start without a proper diagnosis of the original fault and what caused it.
- The only real way to obtain proper preparation is to shotblast completely to a *minimum* grade of $Sa2\frac{1}{2}$ (SSPC-SP10). Witness this activity yourself, as repairs are often done to emergency timescales that do not encourage careful and thorough surface preparation.
- You need to check the new paint system. The best way to do this without wasting time is to talk directly to the paint manufacturers; they can give you a very quick response about the suitability of a paint system for a particular purpose, compatibility of coats, and the details of preparation and application techniques required.
- Witness the application of all three coats if this is practical. It not, make sure that reliable visual and dft checks are done after each coat.
- Repeat all the final checks again after the repainting is finished. This is one area where site rectification of any subsequent defects would not be good practice. Expect some equipment manufacturers to place you under pressure to release the item before you have checked it properly.

Cosmetic painting defects can be treated differently. Unless the substrate material is actually exposed, or you can find evidence of a more serious problem, it is up to you whether or not you feel it is necessary to witness or re-inspect cosmetic repairs. Cosmetic defects

> **FIG 10.3**
> **Job well done: Paint inspections**
>
> 1 For painting, integrity and specification criteria refer to the steps of painting *process*. They are
>   - proper surface preparation
>   - a suitable paint system
>   - correct application.
> 2 The main types of paint are air-drying and two-pack epoxies. They have different properties, thickness and *problems*.
> 3 Proper surface preparation is essential. This *normally* means shotblasting to the standard $Sa2\frac{1}{2}$ (SSPC SP-10) – a well-blasted surface with all the millscale removed.
> 4 You will find it is an advantage to know the content of the technical standards mentioned covering surface preparation and paint testing. It is not necessary to understand the many chemistry-based standards that exist.
> 5 *Visual examination* is the core of effective inspection of paintwork. Learn the common types of painting defects and how to describe them properly.
> 6 The main works tests are for dry film thickness (dft) and adhesion (the 'pull-off' test).
> 7 Major repairs usually require complete shotblasting and repainting: partial solutions are normally ineffective.
> 8 *Maintain your focus.* Don't waste time picking out small cosmetic defects. Look for the big issue.

should be repaired, even if you have concluded that they do not compromise integrity or specification compliance. It is best not to say or report anything, however, that would enable these defects to be misunderstood out of context by a distant client or end user. Do not base a whole source inspection report around lists of cosmetic defects that are easily repaired, not if you want to be taken seriously.

Chapter 11

Inspecting linings

A disproportionate number of problems on petrochemical, desalination and general process plants are caused by the failure of rubber and glass-reinforced polymer (GRP) linings. It seems that these failures tend to happen at the most awkward time, perhaps 6–12 months after plant commissioning. Because linings are used as an alternative to expensive materials (the purpose of linings is to resist corrosive and/or erosive effects of the process fluid), they find use on large vessels and equipment items. This means that downtime and repair costs are inevitably high.

More than 90% of all lining failures can be attributed to poor application. It needs only a single application error to break the seal between an aggressive process fluid and a vulnerable base material. General corrosion, erosion and failure then follow in quite rapid succession; with some process fluids it may take only a few days. For this reason the overriding integrity criterion for linings is the way in which the lining is *applied* to the base material. You will find it useful to apply this simplified view during source inspections to help your focus. It is easily within the capability of a source inspector (SI) to find and eliminate nearly *all* the application errors that cause linings to fail – the techniques are straightforward and there is a set of well-developed and proven tests.

11.1 Rubber linings

Rubber linings are in common use to protect components in sea-water cooling, condenser cleaning and chemical dosing systems, and to resist aggressive process liquors in specialised chemical plants. There are two distinct types of rubber compound used. Natural rubbers are used for general low-temperature sea-water or slurry system applications. Synthetic compounds such as nitryl, butyl or neoprene are used for operating temperatures up to 120°C or when oil is present. Both natural and synthetic rubber formulations can be classified broadly into either a hard or a soft type. Hard rubbers have a higher sulfur content, which in some cases forms hard compounds commonly called *ebonites*. It is mainly used for temperatures up to 100°C. The terms *hard* and *soft* rubber are generally accepted as referring to specific hardness levels.

The most widespread method of application of both natural and synthetic rubber linings is to apply the rubber in sheet form, bonding it to the vessel or component surface with a suitable adhesive such as isocyanate. The rubber sheet is normally 3, 4, 5 or 6 mm thick. This process is done manually; in large vessels the sheets are laid down in overlapping courses whereas for smaller components, such as valves and pipes, smaller pieces are stretched over the component profile, stuck down and the excess is trimmed off. As a final step the lining is vulcanised by heating to approximately 120°C in a steam-heated autoclave or oven for several hours. This develops the final physical and chemical properties of the rubber.

From this description, two points stand out. First, rubber lining is a highly *labour-intensive* activity – it is practically impossible to automate the sheet-lining of engineering components, apart from perhaps small, mass-produced fittings. This means that the whole activity is skill dependent, and subject to human error. Second, rubber lining is an activity that requires careful *process control* because small variations in the way in which application, bonding and vulcanising is done will affect significantly the final result.

Acceptance guarantees

The process of rubber lining is analogous to welding, in that acceptance guarantees in contract specifications normally refer to technical standards, rather than specify directly a separate list of tests and pass/fail criteria. Expect to see only a rubber type, thickness and hardness value stated. In addition, the assessment of lining *workmanship*, with all the subjectivity that it implies, is the responsibility of the SI – and it forms a major part of the implicit requirements of a contract specification.

Special design features

Those parts of a process system that are lined as a means of protection normally incorporate a number of design *features*, incorporated in the recognition of the specific types of problems that arise when a lining has to be applied to a component (and has to stay on). Figure 11.1 shows the details.

FIG 11.1
Rubber-lined components – design features

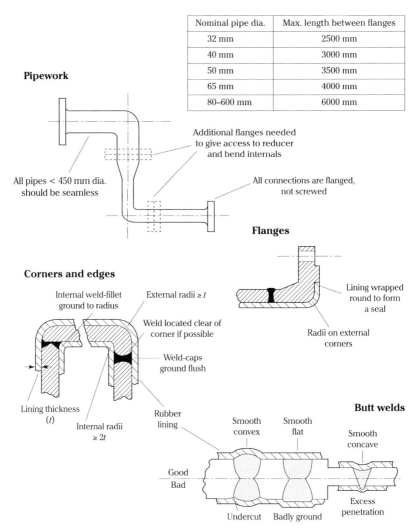

11.2 Specifications and standards

Although the chemical analyses and mechanical properties of rubbers are the subject of vast and comprehensive technical standards, you do not need large volumes of information with you for the purposes of source inspection. Many of the standards concentrate on the chemistry of rubbers and are predominantly of 'laboratory interest' and are therefore not of direct use during a source inspection. There are a small number of standards, however, which *are* of use to an SI.

BS 6374: *Lining of equipment with polymeric materials for the process industries* is an old (1985), but still relevant standard. It has real practical use and is divided into several sections, each dealing with a particular category of lining materials. They are listed below.

- BS 6374: Part 1 covers the application of sheet thermoplastics. The most common linings that you will meet in this category are those with a high resistance to acid. These are commonly used to line process vessels in chemical plant applications.
- BS 6374: Parts 2, 3 and 4 cover the application of non-sheet thermoplastics, stoved thermosetting resins and cold-cured thermosetting resins, respectively. You may also meet these lining materials in chemical plants.
- BS 6374: Part 5 is a specification for lining with rubbers. It shows desirable design features. Some care is needed in implementing these; however, the standard makes clear that it contains guidelines rather than mandatory requirements. There are useful sections on fabrication of vessels intended for lining, common lining defects and relevant rubber properties. It describes tests for rubber hardness and continuity – which are two of the common tests that you will witness frequently during works inspections. This is a simple and well-written standard, which you should find of real practical use.

11.3 ITPs for lined equipment

At best, the ITP for a rubber or GRP-lined vessel component will contain only broad statements of source inspection requirements. Often it merits merely a single line entitled 'lining inspection' towards the end of the ITP. This is a weak and rather misleading interpretation of what is actually required, as small operator or process errors made during the application process have an absolutely key impact on the final integrity and longevity of the lining. Many of these errors are so small as to be very difficult to detect at the final inspection stage using the practical

FIG 11.2
Rubber linings: ITP steps

INSPECTION STEP	COMMENTS
1. Check design features	Check for features which will cause lining problems (BS 6374: Part 5).
2. Fabrication check	Check weld profiles, edge radii and general finish for smoothness.
3. Surface preparation	A minimum grade of Sa $2\frac{1}{2}$ (SSPC-SP-10) is required.
4. Materials check	Check the shelf-life of the unvulcanised rubber sheet and the bonding adhesive – also make a visual examination of the rubber sheet for defects (pores, blisters or tears).
5. Witness lining application	Monitor ambient conditions (at least 3°C above dew point), adhesive and methods of sheet jointing.
6. Pre-vulcanisation inspection	Check the workmanship of the completed lining. Do a continuity (spark) test before fitting cover-straps to seams. Review the vulcanisation procedure – time, temperature (125–160°C) and humidity requirements.
7. Post-vulcanisation inspection	Visual inspection, hardness test, full continuity (spark) test and adhesion (rapping) test.
8. Repairs	Check of local repair procedure and witness of re-tests after further vulcanisation.
9. Pre-shipping inspection	Final check for correct packing to avoid damage to the lining in transit.

test techniques that are available. The message therefore is that SIs should inspect *during* the lining process, armed with a knowledge of the types of errors that can occur. You need to get really close to the process – it is not sufficient only to inspect the lining at the final inspection stage. Such interim inspections are rarely shown on the ITP. Be prepared therefore to apply a little interpretation to what is written, in the knowledge that the body of technical evidence about the root causes of lining failures is largely on your side.

If, as part of your early input to a contract specification, you have the opportunity to influence ITP content, raise rubber linings as an important issue. Try as a minimum to get compliance with a published standard quoted as a specific requirement. Specify interim inspections of the lining preparation and application process in the ITP if you can. Figure 11.2 shows an example of what to include – this gives a much more comprehensive approach and decreases the probability of subsequent lining failures.

11.4 Test procedures and techniques

There is an accepted set of straightforward testing techniques, which are used during final source inspections of natural and synthetic rubber linings. Do not think of them as perfect test techniques; they have a few weak points but are the best tests available for use in a practical source inspection situation.

The visual inspection

It is essential to perform a close and thorough visual inspection, preferably before and after vulcanisation. A general cursory examination is not sufficient, your objective should be to inspect each full or part-sheet that has been used in the lining, and every joint and seam. Use the following guidelines (see Figure 11.3).

- **Use adequate lighting** when making a visual inspection of an internally lined vessel. You need an a.c. lead lamp for this – a torch is not bright enough. In large vessels you should use a wooden ladder to reach the upper walls and top surfaces. Remember that you are making a *close* visual inspection, not just observing from a distance.
- **Adopt a methodical approach**. Start your examination at one end of the vessel and inspect each lining sheet 'course' in turn, working along each seam to the other end. Check each sheet for the following features.
  - Physical damage such as tears, cuts and punctures.
  - Adhesion to the surface (just do this visually for the moment). Look for obvious bulges, blisters or ripples, especially on concave surfaces, that indicate where the sheet is not properly glued down. Check carefully at the location of any set-through nozzles; you will sometimes find air bubbles where the sheet has not been properly smoothed over the internal nozzle-to-shell fillet weld. Pay particular attention to the internal lining of small-diameter pipes where application of the rubber sheet is difficult. One of the best areas to check is on a tight outside radius; for example, where the rubber sheet is lapped over the face of a pipe flange. If you push the lining with the heel of your hand, you should be able to feel whether or not it has adhered properly to the surface. Unfortunately, there is no in-situ quantitative test for adhesive bond strength; the only way is to use test pieces.
  - Smoothness. Sight along the sheet with the light behind you and

FIG 11.3
Visual inspection of rubber-lined vessels

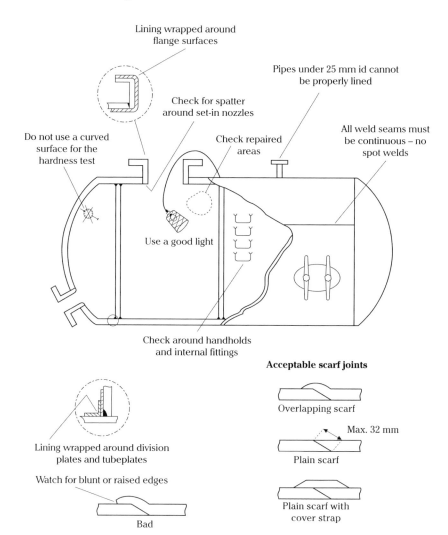

look for evidence of any weld spatter or small foreign bodies left underneath the sheet.
– Seams. Check the scarf joints or overlaps where the lining sheets have been joined. For a typical 4 mm thick rubber lining there should be at least 16 mm (four times the sheet thickness) of surface contact between adjoining sheets, but not normally more than 32 mm. Figure 11.3 shows the three main types of scarf joint that you will see. Run your hand along each seam to make sure the edge is firmly stuck down. You should not be able to feel a loose edge when running your fingers against the lap directions and the scarf edge should be sharply feathered to give a good tight joint. A raised 'butt' on the scarf edge is bad practice.
– Nozzles and flanges. The lining sheet should be correctly wrapped round all nozzles and flanges to provide a complete seal against the process fluid. Pay particular attention to small fittings of less than 30 mm diameter because it is more difficult to get the lining to adhere to their tight internal radii.
– Internal fittings. Fittings such as baffles, separators and internal tubeplates are often designed as permanent fixtures in a lined vessel to try to avoid practical lining difficulties. If you encounter removable fittings, make sure that the lining sheets have been applied in a way that enables a full seal to be made. If separate cover strips of lining have been used, check they are well jointed and stuck down. Do not forget miscellaneous fittings such as handholds, brackets and manhole door pivots.

As with other types of visual inspection, it is useful to use a checklist. This will ensure that nothing is missed, and form a useful addition to your inspection report. One way to develop this checklist is to go through BS 6374, listing the features it mentions, then sort them into a logical order for use during your visual inspection.

Rubber hardness check

The hardness of a rubber lining is a good measure of abrasion resistance, and an indicator of whether the vulcanisation process has been completed correctly. Post-vulcanisation hardness is therefore one of the main specified properties. Hardness is measured using a simple hand-held indentation meter. The tip is pushed into the rubber and the hardness reading is shown on a spring-loaded dial. There are two main hardness scales in common use. Soft rubbers use the International Rubber Hardness Degree (IRHD) scale – this goes from 0 to 100 degrees and is roughly proportional to the Young's modulus of the rubber. A soft rubber is generally considered as having a hardness of between 40 and 80 degrees IRHD (below 40 degrees is a very soft rubber and these are used less often). Hard rubbers range from 80 to 100 degrees IRHD *or* may be referred to the 'Shore D' scale. For most engineering applications, hard rubbers are considered as those in the range 60–80 Shore D degrees. There is a generally accepted tolerance of +5 degrees allowed for both IRHD and Shore D measurements. It is important to check hardness readings at several points around the vessel lining to obtain the average reading. Use the following general guidelines.

- Always take readings on a flat surface. Measurements from curved surfaces can introduce errors of 10–20 degrees IRHD.
- Take the readings at the centre of the lining sheets or at about 1 m apart. To obtain representative results, it is best to take at least three readings per individual rubber sheet.
- Concentrate on those regions of the vessel that were in the *lower* position when the vessel was placed in the autoclave for vulcanisation (obtain this information from the lining contractor). You will find that these lower areas are sometimes softer due to condensation accumulating from the steam heating, preventing the vulcanisation from being completed properly. You may see some evidence of this during your visual checks – poorly vulcanised areas will be a noticeably different shade (normally lighter) than the rest of the lining.
- The ambient temperature will have some effect on the hardness readings that are taken. Try to ensure that the local temperature inside the vessel is between 15°C and 25°C. If the vessel is placed in strong sunlight, the internal temperature will rise and give misleading readings.
- Check the calibration of the hardness tester before use. Calibration test pieces should be kept with each tester.

Spark (holiday) testing

The purpose of spark testing is to check for *continuity* of the lining. This is quite a searching technique; an electrical spark will locate a pinhole or small discontinuity in a lining sheet that would not be detected visually. It will not detect any errors in adhesion. A high-frequency, high-voltage a.c. supply (it is common to use a minimum of 20 kV) is applied between the parent metal of the vessel and a hand-held probe. The probe is then passed just above the surface of the rubber. A strong spark will jump the gap when there is a conducting path caused by a pinhole or discontinuity. There are a few points to note.

- It is essential that the voltage be sufficient to enable the spark to jump the maximum air path that it is likely to encounter. This would be the path through a faulty scarf joint. For a typical 4 mm thick lining, it is approximately 32 mm. You can do a test by simply checking that the spark will jump this length of gap from the probe to an unlined area such as a bolt hole.
- Do not confuse stray 'air path sparking' from the probe with the large blue spark that will jump to a real leakage path. There will always be some small streamer sparks into the air, particularly in damp conditions.
- Dim lighting makes it easier to see the sparks.
- Be methodical in your approach to spark testing by working round all scarfed seams, nozzle joints and any cover straps over fittings. Check around manholes and handholes. Test the bottom surface of the vessel where physical damage is more likely, and in difficult locations where accurate seams are more difficult to make.

For large, rubber-lined items such as vessels and ductwork, it is common to perform a spark test both before and after vulcanisation. This is also good practice for vessels which are lined and vulcanised at the construction site, under non-ideal working conditions. For seams which are fitted with cover straps, a preliminary spark test should be done before the cover straps are glued on.

Adhesion tests

Poor adhesion is a common cause of failure of rubber linings; the lining sheet can peel off the base material, often starting from an air bubble or loose radius, until a seam is reached. The process fluid then gains access underneath the lining and general corrosion and failure happen within a short time.

Under laboratory conditions it is possible to obtain adhesive bonding between rubber and metal that is stronger than the rubber itself (specialised laboratory-based standards cover this). In practice, variables such as the adhesive mix, ambient conditions and operator skill make this difficult to achieve, particularly around internal radii, sharp corners and fillet weld profiles. A crude but reasonably effective works check on adhesion can be carried out by doing a *rapping test*. The rubber is rapped using a special ball-headed hammer. Areas of good adhesion will give a firm ringing sound. If there is an air gap between the lining sheet and base material, the sound will be duller. This is by no means an exact test; it is easy to disagree on the comparison of sound obtained. Small variations of hardness of the lining can also cause differences. The test will also only identify areas where there is significant lack of adhesion leading to an air gap. It will not detect low strength or partial adhesion, both of which are root causes of many lining failures, nor will it detect small patches of poor adhesion on scarfed seams. The rapping test *is* a useful and practical site test, as long as you recognise its limitations. It is certainly not a substitute for witnessing the lining activity (see the recommended ITP steps in Figure 11.2), and making careful checks of surface preparation, adhesive mix and application technique.

11.5 Metallic linings

The term *metallic lining* is used to encompass a large number of processes. These range from cladding base material by welding on sheets of corrosion-resistant alloy to more complex processes where a very thin layer of material is added to a metal surface by deposition, spraying or electrolytic action.

Integrity criteria: metallic linings

As with rubber, the main purpose of most metallic linings is to prevent corrosion and erosion of the base material by an aggressive process fluid or environment. The main criterion therefore is *integrity* – the lining must provide a perfect seal if it is to be effective. This seal is provided by welding in the case of loose-clad components and a metal-to-metal bonding in the case of sprayed, dipped or electrolytically applied metallic coatings.

It is safe to say that, *in general*, metallic linings involve less source inspection activities than do rubber linings. One reason is that the processes themselves tend to be more technologically complex and need to be quite closely controlled if they are to work at all. Metal spraying and electrolytic plating are good examples of this. A second reason is that the inherently simpler processes, such as cladding, come under the control of accepted welding procedures and practices. This is a reasonably well-controlled regime, as we saw in Chapter 6.

We can look briefly at the two most common metallic lining

techniques that you will meet in a works inspection situation: loose cladding and galvanising.

Loose cladding

Loose cladding is most commonly used on fluid and process systems where there are aggressive process conditions, such as in chemical or desalination plants. The cladding material is usually either high-nickel alloy such as Inconel, or Monel, or stainless steel. The term 'loose cladding' means that the lining sheets are not bonded to the base material (usually low-carbon steel) over their complete surface area and attachment is achieved by fillet-welding around the periphery of the lining sheets. This may be supplemented by plug welding, which is a MIG welding technique in which spot welds penetrate through the cladding sheets into the base material at regular intervals. ASME VIII has a section dealing specifically with clad vessels.

Some particular guidelines relevant to source inspections are listed below.

- **Design features**. ASME VIII contains guidelines on desirable design features for clad vessels. You can do a broad design check using the construction drawings if you are inspecting conventional vessels, but other equipment is more difficult.
- **Welding**. The welding between the cladding sheets and base material should be covered by the system of WPSs, PQRs and welder qualifications described in Chapter 6. It is good practice to carry out surface crack detection on these welds, applying the same standards as you would for other dissimilar material welds.
- **Pneumatic testing**. Some clad components are subject to a pneumatic test. Air is introduced into the space between the parent material and the cladding, then soap solution is used to check for leaks.
- **Surface preparation**. It is good practice to shotblast the inside surface of the base material before cladding. This removes any active corrosion products and minimises subsequent deterioration. The inside surface of the lining sheet is normally mechanically cleaned before fixing.
- **Final inspection**. It is worth making a final inspection of loose-clad components. Check that all the seal welding is properly completed. Small fittings and fasteners should be of the same material as the lining, to prevent galvanic corrosion in use.

Figure 11.4 shows some specific checks on common loose-clad components.

FIG 11.4
Inspecting loose clad components

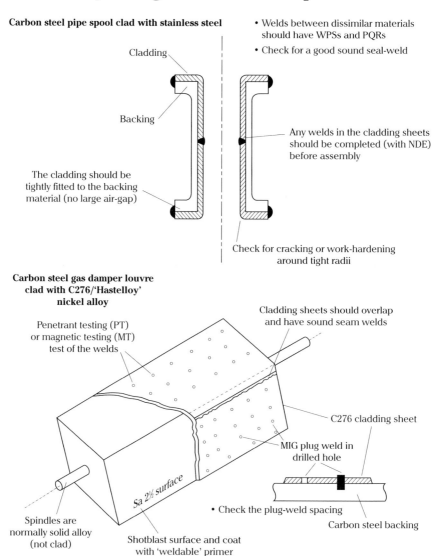

Galvanising

Galvanising is a generic term for the coating of iron and steel components with zinc. Its main use is to protect the base material against attack from a corrosive atmosphere or water. It is predominantly used externally on engineering components rather than to protect internal surfaces and can be used instead of painting or occasionally in conjunction with it. Much of the benefit in using zinc lies with its position in the periodic table; zinc is anodic to steel and will therefore catholically protect it, even if the zinc layer suffers scratching or damage.

Galvanising is unlike many types of cladding or coating because the zinc forms a *chemical* bond with the iron in the surface layer of the substrate. Purer zinc then bonds with the iron/zinc compound until the surface layer is almost pure zinc. Several different processes are often misleadingly termed 'galvanising'. Processes such as *sherardising* (coating items using a zinc-rich dust), *zinc plating* (actually an electro-deposition process) and *zinc spraying* tend to be used for small, mass-produced items such as fasteners, fittings and precision components. The most common process for source inspection is the technique of *hot dip* galvanising. This is widely used for structural steelwork and large fabrications that are exposed to the atmosphere. Hot dip galvanising is a relatively low-technology process in which the component is cleaned and then dipped in a bath of molten zinc. The zinc compounds form on the surface and remain when the component is removed from the bath.

The most widely used technical standards for hot dip galvanising are

- **ASTM A123:** *Standard specification for zinc coatings on iron and steel products*
- **EN ISO 1461:** *Hot dip galvanized coatings on iron and steel articles – specifications and test methods*

These are good general guides which contain most of the information that you need during a source inspection.

The visual inspection

Although hot dip galvanising can be considered a proven and reliable process, errors can still be caused by incorrect composition of the zinc bath, or by poor preparation of the parent metal. A good visual inspection is therefore important. Key points to check are listed below.

- **Surface appearance**. If the base metal has an even surface, then the

surface of the zinc coating should also have a smooth and even finish. Uneven features such as weld laps and seams will show clearly through the coating; normally this is a purely cosmetic problem.
- **Colour**. It should be bright. The only common exception is for some types of high-silicon steels in which the alloying elements cause the coating to be slightly dull. Check the steel type if you see this, to make sure that this is the reason and that it is not a problem with the purity of the molten zinc mix.
- **Staining**. Technical specifications (and good practice) call for the galvanised surface to be free of staining. The most common type of staining is caused by storing the galvanised components in wet conditions for long periods, producing an effect which looks rather like white rust. If necessary the components can be subjected to a phosphating or chromating surface treatment after galvanising to stop this happening.
- **Coating discontinuities**. These are more often the result of mechanical damage than problems with the chemistry of the coating process. Look carefully for this around exposed edges and corners. Small areas up to 30–40 mm^2 in size can be repaired (using a low-melting-point zinc filler rod, applied rather like solder), but this is not good practice for large areas.

Coating weight test

It is a convention for galvanising specifications to specify the coating *weight*, expressed in g/m^2, that has to be applied. The most accurate way of checking this is by a *stripping test* in which a small test piece is immersed in an acid solution and the weight of coating removed (by dissolving) is measured. Practically, the most common method used during source inspection is a simpler weight difference method – the component is weighed before and after galvanising and the difference is divided by the surface area to give a g/m^2 coverage figure. Figure 11.5 shows typical coating weights suitable for various components. Note that a rough (shotblasted) surface of the parent metal will accept a thicker coating than a smooth one.

Coating uniformity (Preece) test

The uniformity test is used as a complement to the weight test as it is not sufficiently accurate to use alone as a test of coating thickness. The purpose is to check the 'evenness' of the coating, normally written as a desirable criterion in most galvanising contract specifications. The test is

FIG 11.5
Hot dip galvanising: Weight and uniformity tests

COATING WEIGHT

Parent material	Minimum galvanised coating weight
Steel 1–2 mm thick	335 g/m^2
Steel 2–5 mm thick	460 g/m^2
Steel > 5 mm thick	610 g/m^2
Castings	610 g/m^2

An approximate conversion from coating weight to coating thickness is:
1 g/m^2 ~ 0.14 µm

Coating weight can be determined by one of the following.
- An acid 'stripping' test (the most accurate 'laboratory' method).
- Direct measurement of the weight difference of the component before and after galvanising.
- Sometimes a simple magnetic or electronic thickness measuring device is used to calculate an approximation of the coating weight.

COATING UNIFORMITY
($CuSO_4$ or 'Preece' test)
The uniformity test is complementary to the coating weight check. The steps are listed below.
- Prepare test specimens of the galvanised material.
- Dip the specimens four times successively into a $CuSO_4$ reagent solution – each dip should last 60 s and the specimens must be rinsed between dips.
- If a permanent film of red-brown metallic copper appears on the surface of the metal, the coating has failed the uniformity test.

> **FIG 11.6**
> **Job well done: Rubber linings**
>
> 1. The root cause of most rubber lining failures is poor application – you can help the situation by making *interim* inspections.
> 2. There are a number of small, but important, design features that are desirable for rubber-lined vessels and components. It is worth checking these.
> 3. Common techniques are used to test for:
>   - hardness
>   - continuity (the spark test)
>   - adhesion (the rapping test).
>
>   Always do a close visual examination – there are several specific things to look for.
> 4. Defects in rubber lining *can* be repaired, but you should monitor the procedure to ensure it is controlled closely.
> 5. For metallic linings, the main FFP criterion is *integrity*, so that the base material is not exposed.
> 6. Loose cladding is primarily a welding-based process. You can use some of the information in Chapter 6 for guidance.
> 7. Galvanising (coating steel by dipping in molten zinc) is a common process. One of the main tests is to check the *weight* of the zinc coating.

commonly called a *Preece test* (ASTM A239) and consists of dipping a test piece into a copper sulfate solution. The solution will expose the base metal in any areas of thin coating and deposit a red-brown layer of metallic copper. This is an indication that the coating has weak areas and is therefore unacceptable.

In practice, most works inspections of galvanised components are performed on a small sample chosen from what may be a large batch of similar items. Accept this as convention, but do not neglect a general visual inspection of the whole batch, just to make sure.

FIG 11.7
Watch out for: Rubber linings

Non-conformance	Corrective action
Poor surface preparation	Don't accept poor surface preparation – you should ask for it to be done again. Check that: • Weld caps are properly smoothed off (they don't necessarily have to be ground totally flat). • Shotblasting is to the level SIS 055900 Sa $2\frac{1}{2}$. There should be no significant scale or rust and the surface should be a light, speckled grey.
Wrong rubber lining thickness	There is a general allowable tolerance of $+10\%$ on the (unvulcanised) thickness of the rubber sheet. If the sheet is too thick, this is unlikely to cause any harm. Accept it. If the sheet is more than 1 mm too thin, reject it. The vessel needs stripping, shotblasting and re-lining.
Defects found before vulcanisation	Local repairs are possible. The solution is to cut out the defective area and insert a repair piece. Use scarf joints with a 10–15 mm overlap. Make sure the new seams are very well stuck down and all air bubbles are excluded.
Incorrect hardness results	First check the allowable tolerances in BS 903. A tolerance of $+5\%$ is acceptable on any individual reading. Hardness is usually specified as a *minimum* value required. The most likely cause of low hardness is incorrect vulcanisation – it is acceptable to re-vulcanise the component. If this does not improve the situation, it is almost certainly the wrong grade of rubber. Discrepancies of less than 10 degrees (Shore D or IRHD) can often be accepted under concession – more than this and the lining cannot really be considered fit for purpose.
Defects found after vulcanisation (by spark tests)	Again, repairs are possible. After repair, the area should be locally vulcanised using special heated 'irons' or similar. Repeat the spark test, hardness measurements and rapping test on the repaired area.
Poor seam overlaps or short scarf lengths	Poor scarf joints can be overlain with cover strips – separate strips of lining approximately 70–100 mm wide. This is acceptable to a limited extent but is not a replacement for neatly scarfed seams. Cover straps may peel off sharp corners or tight inside radii in use. It is best to use them sparingly.
Lack of (or poor) adhesion	Patches of poor adhesion are usually symptomatic of a more general problem with the adhesive or surface preparation. Always assume that the situation will get quickly worse in use. Except for very small areas near the centre of lining sheets, poor adhesion is unacceptable. Issue a NCR and ask for it to be done again.

Part B

API SIFE EXAM PREPARATION

Chapter 12

The API Individual Certificate Programmes (ICP)

API certificate examinations

API certificate examinations come under the general banner of the API Individual Certificate Programmes (ICP). This is an expanding suite of examinations servicing the inspection/integrity part of the engineering industry.

Note these general features of these API (ICP) exams.

- API exams are **not for beginners**. They are not really aimed at trainees or new entrants to industry.
- API impose **entry requirements** for candidates wishing to register for many of the examinations.
- **API examinations are difficult** – owing mainly to the long scope of the published body of knowledge (BoK) that forms the source material for each exam category. For some this can amount to 1500+ pages of published codes and recommended practice (RP) materials.
- API examinations (as their name suggests) are written in US style, based on US codes, practices and examination style. This is consistent in itself, but can differ significantly from that used in other parts of the world. This is an important point which (depending on your background) may have a real effect on your ability to understand the programme material and pass the exam.

The API ICP scope

The list below shows the full scope of the API ICP examinations for 2016. At any time the current list of ICP programmes is shown on www.api.org/certification-programs. The content of this book fits in with the source inspector fixed equipment (SIFE) scope.

- API 510 – Pressure vessel inspector

- API 570 – Piping inspector
- API 653 – Aboveground storage tanks inspector
- API TES – Tank entry supervisor
- API 1169 – Pipeline inspector
- API 571 – Corrosion and materials professional
- API 577 – Welding inspection and metallurgy professional
- API 580 – Risk based inspection professional
- API 936 – Refractory personnel
- API SIFE – Source inspector fixed equipment
- API SIRE – Source inspector rotating equipment
- API QUTE – Qualification of ultrasonic testing examiners (detection)
- API QUPA – Qualification of ultrasonic testing examiners (phased array)
- API QUSE – Qualification of ultrasonic testing examiners (sizing)
- API IA-Q1 – Internal auditor Q1
- API IA-Q2 – Internal auditor Q2
- API A-Q1 – Auditor Q1
- API A-Q2 – Auditor Q2
- API LA-Q1 – Lead auditor Q1
- API LA-Q2 – Lead auditor Q2

The SIFE programme scope

The BoK set by API for the SIFE examination is large, diverse and widespread. It uses codes and published documents from five different organisations

- American Petroleum Institute (API)
- American Society of Mechanical Engineers (ASME)
- American Welding Society (AWS)
- American Society of Non-destructive Testing (ASNT)
- Society for Protective Coatings (SSPC).

As you would expect, these documents have different structure, content and style. There is a fixed package of code documents for the SIFE examination. This is shown in the 'code effectivity list' in Figure 12.1. These represent the BoK for the subject as decided by API.

Registering for the API SIFE exam

This is a completely separate activity to participating in any training programme. It is each candidate's individual responsibility to pay and

FIG 12.1
The API SIFE 'Code effectivity list'

API American Petroleum Institute
- **Source Inspector Exam Study Guide**
- **API Recommended Practice 572**, Inspection of Pressure Vessels, 3rd Edition, November 2009, Sections 3 and 4
- **API Recommended Practice 577**, Welding Inspection and Metallurgy, 1st Edition, October 2004
- **API Recommended Practice 578**, Material Verification Program for Alloy Piping Systems, 2nd Edition, March 2010
- **API Standard 598**, Valve Inspection and Testing, 9th Edition, September 2009

AWS American Welding Society
- **D1.1** Structural Welding Code – Steel, 22nd Edition, March 2010 with Errata
- **WI** Welding Inspector Handbook, 3rd Edition, November 2000

ASNT American Society of Non-destructive Testing
- **Recommended Practice SNT TC-1A** Personal Qualification and Certification in Non-destructive Testing Personnel, 2011 Edition

ASME American Society of Mechanical Engineers, Boiler and Pressure Vessel code (BPVC), 2010 with 2011 Addendum, July 2011

Section II Materials, Part A:
- Sections SA-20, SA-370, SA-6

Section V Non-destructive Examination:
- All destructive in Subsection A, Article 1, Appendix 1 and Subsection B, Article 30, SE-1316
- Articles 1, 4, 6, 7, 9, 10 and 23 (section 797 only)

Section VIII Rules for Construction of Pressure Vessels, Division 1:
- All definitions in Appendix 3
- Sections UG 4–15; UG 75–85; UG 90–103; UG 115–120
- Sections UW 1–3; UW 5; UW 26–42; UW 46–54; UW 60
- UCS 56–57

Section IX Welding and Brazing Qualifications, welding only:
- QW 100–190; QW 200–290; QW 300–380
- QW 400–490; QW 500–540

ASME American Society of Mechanical Engineers

B31.3 Process Piping, 2012 Edition
- Chapter I, III, IV, V, VI

B16.5 Pipe Flanges and Flanged Fittings, 2009 Edition
- Chapters 1–8

SSPC Society for Protective Coatings

SSPC – PA 2 Procedure for Determining Conformance to Dry Coating Thickness Requirements, May 2012

SSPC Surface Preparation Guide, the following sections only:
- SSPC-SP1 Solvent Cleaning, 2004
- SSPC-SP3 Power Tool Cleaning, 2004
- SSPC-SP5 NACE 1 White Metal Blast Cleaning, 2007
- SSPC-SP6 NACE 3 Commercial Blast Cleaning, 2007
- SSPC-SP7 NACE 4 Brush-Off Blast Cleaning, 2007
- SSPC-SP10 NACE 2 Near-White Blast Cleaning, 2007
- SSPC-SP11 Power Tool Cleaning to Bare Metal, 2012

Note: The current version of effectivity lists for all API ICP programmes is available on www.api.org/certification-programs

register for the API examination for the exam windows scheduled throughout the year. This is done via the websites of API (www.api.org) and their exam site contractor, Prometric (www.prometric.com).

Our EDIF group training courses

There is no compulsory training required for candidates who wish to sit for any of the API ICP exams. In theory, to become certified as an API certified inspector, all you have to do is apply to API, meet their acceptance criteria, book your exam (lasting between 3 and 7 hours depending on which ICP you are attempting) and then pass it. Some candidates can pass like this but, for most, unless you have full familiarity with the relevant codes (up to 1500+ pages' worth for some ICPs), you are unlikely to pass the exam and will need to prepare for the examination by enrolling on a training course. The training course will teach you about the subject matter covered, test you using mock exams and so on, and prepare you to take the API exam.

The EDIF Group provide training courses (public and in-house) on a wide range of API, ASME and other courses. You can see these on our website www.edifgroup.com.

Important note: The role of API

API have not sponsored, participated or been involved in the compilation of this book in any way. API do not issue past ICP examination papers, or details of their question banks to any training provider, anywhere. API codes are published documents which anyone is allowed to interpret in any way they wish. Our interpretations in Part B of this book are built up from a 15-year record of running successful API training programmes in which we have achieved a first-time pass rate of 90%+. It is worth noting that most training providers either do not know what their delegates' pass rate is, or don't publish it if they do. API sometimes publish pass rate statistics – you can check their website www.api.org and see if they do, and what they are.

Chapter 13

The API SIFE exam questions; what to expect

The API SIFE examination consists of 100 multiple choice questions, all of which have to be answered closed book. API do not always quote what the required pass mark is, but it is normally around 70%.

13.1 Exam question format: what to expect

API ICP exam questions are compiled using a system of terminology, grammar, construction and logic that, although not to everyone's taste, is consistent enough in itself. In common with multiple choice examinations worldwide, API questions are compiled to fulfil two main purposes

- to test your simple recall of facts, statements and opinions expressed in the code documents
- to test your skills of analysis and evaluation when faced with closely spaced answer options, based on a question that contains limited information about the historical context of the equipment or situation.

The difficulty lies with the high number of SIFE-related documents that are covered – the codes are long documents containing lots of technical information and potentially several thousand exam question opportunities.

13.2 Some hidden secrets about API exam questions

Knowingly or unknowingly (it doesn't matter which), API ICP exam questions fall into four distinct types; see Figure 13.1. The exam question bank is built up from an informal selection of 'parent'

FIG 13.1
Four types of API exam questions

TYPE 1 QUESTIONS:

Verbatim wording quote

Q. The stresses imposed on a vessel during a pressure test are effectively?

Ans. *Static*

TYPE 2 QUESTIONS:

Based on loose paraphrase or 'intent'

Q. The stresses imposed on a vessel during a pressure test ignore?

Ans. *Fatigue conditions*

QUESTIONS ARE BASED ON SOURCE MATERIAL APPEARING SOMEWHERE IN THE CODES… LET'S SAY THIS IS IT:

The *objective* of pressure tests is sometimes misunderstood. It is part of the system of verifying the integrity of a vessel but it has its limitations. The stresses imposed on a vessel during a pressure test are effectively static; they impose principal stresses and their resultant principal strains on the vessel. This means that what they test is the resistance of the vessel only to the principal stress and strain fields, not its resistance to cyclic loadings (that cause fatigue), creep or the other mechanisms that have been shown to cause vessels to fail. Hence the pressure test is *not* a full test of whether the vessel will fail as a result of being exposed to its working environment and the incidence of steel vessels actually failing catastrophically under a works pressure test is quite small . A pressure test **is not** a 'proving test' for vessels that have not been properly checked for defects (particularly weld defects). It is also not a proving test for vessels where unacceptable defects have been found – so that the vessel can be somehow shown to demonstrate integrity, in spite of the defects.

TYPE 3 QUESTIONS:

Elimination or 'least wrong' answers

Q. A pressure test on a vessel is?

(a) Unlikely to result in failure
(b) A test for fitness-for-purpose
(c) A test for all defects
(d) A proving test

Ans. (a) It may not always be correct in all contexts, but it is the *least wrong* based on the code text it was sourced from.

TYPE 4 QUESTIONS:

Based on general knowledge that may not be directly traceable to the form of words in the code.

Q. Pneumatic tests are?

Ans. *Performed in a blast-pit or underwater for safety reasons*

questions of each type, each one then being supplemented by two or three 'brother' questions to bulk up the number of questions available in the bank. These are similar in style, logic and construction to their parent questions, but with some change introduced to affect the result, sense or semantic form of the answer. Essentially they are the same question, just in simple disguise.

The question set for each individual exam date is then chosen from the question bank, with some crude screening to ensure that too many parents and brothers do not appear in a single exam paper. This avoids the risk of leaving the following exam short of particular topics, with the risk of changing the knowledge profile of the overall exam.

Overall, you can expect all the four question types shown in Figure 13.1 to be represented in the quasi-random question selection as described. There are hidden boundaries as to how many of each will appear in the mix. This results from the nature of the subject, its technical content and whether it is constructed mainly of hard technical facts, experienced-based opinion, or a collection of fluffy ideas revolving around a few core concepts which it is possible to agree are generally a 'good idea'. These boundaries install themselves unannounced, but are not difficult to spot if you look for them.

All these four types are fairly frequent in their use of distractors, negative statements and plausible-sounding incorrect answers that represent common mistakes or misapprehensions.

Type 1: Direct quote questions

Type 1 questions have their origin in the verbatim (word-for-word) form of the code documents. One of the answer options (the correct one) contains a verbatim string of words (most commonly between four and six of them, without gaps between them), exactly as they appear in the code. The other (incorrect) answer options can sound plausible, but are considered incorrect because they either

- contain an incorrect word
- contain a negative (i.e. *not*) to change the sense of the answer
- play on your misreading of such statements such as *higher than* versus *lower than*, *minimum* versus *maximum*, or some similar factual presentation.

In writing type 1 questions, the question setter may be more concerned with the fine semantics of the words that make up the question than ensuring that the stated 'correct' answer is correct in all engineering scenarios, or that the other options are *incorrect* in all engineering scenarios. It is fact (but of course not a necessity) that type 1 questions can be written by question setters with little or no detailed knowledge of the practical engineering scenarios that are possible (but unstated) within the boundaries of the words in the question.

Here is an example of a type 1 question

Q. Testing does not refer to:

(a) Pressure testing techniques
(b) NDE techniques
(c) Non-destructive hardness testing
(d) Tensile strength test techniques

The answer is (b).

Depending on the context it is logically arguable that any of the 'incorrect' options (a), (c) and (d) could be seen as correct by some people. In addition, many people may refer to NDE as 'testing' in everyday discussions. From an exam viewpoint, however, the only possible answer to this question is (b), purely because it is a direct quote from a code definition API 572(3.1.3c). That is why (b) is the answer.

Type 2: Loose quote/paraphrase questions

These are just type 1 questions without the semantic finesse. The 'correct' answer has been chosen from the general meaning of some string of words as they appear in the code, but for some reason has lost the link with the direct verbatim presentation. The reason for the difference may be hazy, and owe more to accident than design. Type 2 questions are tentatively more suitable to API SIFE closed-book rather than open-book questions used in other API exams. If you are a loose thinker, you can be caught out by these questions; watch out for the following words and phrases, which are inserted to catch you out

- *generally*
- *mainly*
- *on balance*.

Questions based on definitions and terminologies often appear as type 2 'brother' questions. They are really better suited to open-book questions, so the answer options have to be widened to make them more reasonable to answer closed-book examinations such as API SIFE.

Here is an example of a type 2 question, and a 'brother' question developed directly from it.

Q. A near-white metal blast cleaned surface is generally:

(a) Coated with millscale
(b) Slightly stained
(c) A near-white colour
(d) Produced by power tools

The answer is (b).

This question is sourced from the joint surface preparation standard SSPC-SP10 *Near-white metal blast cleaning*. Here is an extract of the relevant definitions in the standard.

- *A near-white metal blast surface when viewed without magnification shall be free of all visible oil, dust, dirt, mill scale, rust, coating, oxides, corrosion products and other foreign matter. Random staining shall be limited to no more than 5% of each unit area of surface.*

Note how the chosen answer '(b) *Slightly stained*' has been paraphrased out by this paragraph. The paragraph does not say a near-white specimen *must* be slightly stained, just that a certain amount of staining is allowed. To the question setter the answer is clear, even if the support of the actual wording is less than perfect. Welcome to type 2 questions.

Now here is a type 2 'brother' question developed directly from the one above.

Q. A near-white metal blast cleaned surface is not:

(a) Produced by blasting
(b) Coated in small areas of oxide
(c) A near-white colour
(d) Specified in SSPC-SP10

The answer is now (c).

Look at what has happened here, answer (c) is the same, and in the same place, as it was in its previous parent question. Adding the word *not* into the question is what now makes answer (c) correct, as a near-white surface on blasted steel is not, naturally, actually a white colour. Working backwards from this the question setter now has to find, from the same source paragraph, three statements that *are* true about a near white-metal blast cleaned surface. As the supply of suitable word groupings is very limited, then loose associations often have to do. Now you can see the reasons behind the use of paraphrase (or sloppy wording – call it what you will). Looking at the origin of incorrect answer options (a), (b) and (d) gives us the following.

- Option (a): *Produced by blasting* – a weak choice, a blast surface is produced by blasting (look at the question).
- Option (b): *Coated in small areas of oxide* – true enough, the source paragraph says it has to be free of all oxide.
- Option (d): *Specified in SSPC-SP10* – this one has moved slightly outside the box in assuming that you recognise this as an SSPC-SP10 sourced question. It is either a conscious attempt to see if you do know, or the question setter has forgotten that you may or may not know it, because SSPC-SP10 is not cited in the question text. Either way, this is the kind of thing to expect.

Type 3: Elimination questions

These are a fairly crude attempt to test your powers of reasoning and analysis. Of the answer options, the one that is designated as 'correct' is deemed so because it is **less wrong** than the others. Note that, from a strict technical viewpoint, it does not have to be correct in all engineering or historical circumstances, so the questions are very general, and lack much qualifying information.

Background clues to the correct answer may be found in oblique references in code clauses (often in several sections), but the wording is rarely direct. It is more common for the 'wrong' answers to be justified by an opposite or contradictory statement buried somewhere in the code paragraphs, again probably in several places.

These questions do not sit well with most engineers, because *least wrong*, does not necessarily mean *right*.

Here is an example of a type 3 question.

Q. Ductility is:

(a) Related to material strength

(b) Elasticity
(c) The ability of a material to be formed
(d) Linked to toughness

The answer is (d).

Explanation
The origin of the answer comes from the fact that the ductility of a metal is what provides blunting at crack tips when a crack is trying to propagate, hence increasing the resistance of the metal to crack propagation (i.e. its *toughness*).

Note how the wording of the question and answer option (d) are not intimately matched to this explanation, but are not fully untrue. The other options (a), (b) and (c) have been generated (fairly randomly in this case and without code-related back-up wording) for their partial untruth, although they may look convincing at first glance.

Option (c) has a sprinkling of truth about it, but the ability to be formed is better described as *malleability,* rather than ductility.

Type 4: 'General knowledge' questions

Thankfully, these are fairly rare, as API SIFE examination questions are sourced from the content of the code documents in the BoK. Occasionally, however, some facts are considered sufficiently widely known to be considered general knowledge that an inspector should possess, by virtue of past training and experience. Once again they may be justified by oblique or general statements in the code documents.

Type 4 questions are included to trip up candidates with very poor experience, and should not prove difficult at all to anyone with a reasonable level of experience and background knowledge in engineering subjects. Here is an example.

Q. During new manufacture of fabricated equipment, most defects are found in:

(a) Welds
(b) Forgings
(c) Tubes
(d) Nozzles

The answer is (a). It's all very simple – the question setter has decided that *welds* are where most defects will be found, so thinks you should believe that also. There's even a clue (look – it says *fabricated* in the question). You need to accept that this is the style of some of the questions. No-one has anticipated that you might argue that forgings can contain welds, as of course do some nozzles and tubes, or that it is perfectly possible to have a welded, forged tube made into a nozzle. You are thinking too deeply – this is a valid general knowledge question and the answer is *welds*. It may even be true.

Now, the distractors

All the four question types can be changed (and 'brother' questions created) by the application of a few *distractors*. There are not many of these and they are well known. They also work well; historically, distractors have a 25–40% efficiency at tripping up candidates who have weak thought patterns, rely heavily on rough guessing or are just mentally lazy (even though they think they are trying hard). The main types are discussed below.

Distractor A: The '*not* question'

Negative or 'not' statements are heavily used as distractors in API exam questions. They appear applied to all four question types and in more than 80% of cases the 'negative' appears in the question statements rather than the answer options (where it is a little less efficient at catching you out). Expect 10–15% of API exam questions to have this type of distractor included. Expect a few double-negative questions also (direct or implied). These are difficult for candidates reading them in a second language.

Distractor B: The '*sounds correct*' answers

These are plausible-sounding incorrect answers that either represent common mistakes or misapprehensions, or just simply sound correct, but are not. They catch out guessers, skim-readers and lazy people.

Q. Hydrostatic pressure tests, if repeated several times on a vessel:

(a) Are not a test of fitness-for-purpose in service
(b) Are better than a single hydrostatic test
(c) Are a low safety risk
(d) Can cause fatigue

What would you say the answer is? It is there, hidden behind three distractors and a bit of technical opinion. Option (d) *Can cause fatigue* catches the eye first – you know fatigue is bad, and your choice is fed by the *several times* mentioned in the question. It's not correct, however, because many, many pressure tests would be necessary to produce any significant fatigue regime. Next up, option (c) *Low safety risk* – sounds correct, your mind (how wonderful it is at filling in the blanks) tells you that hydrostatic tests produce lower risk than a pneumatic test. Absolutely true, but that's not the question, so it is not the answer. Grasping at straws, option (b) *Are better than a single hydrostatic test* pops up next. Several tests *have to be better* than one you tell yourself – if you don't find a leak on the first test, you may find it on the second or third. Popular fantasy I'm afraid, if there's a leak you will find it the first time, which is why codes only require a single test. The correct answer is (a). Hydrostatic tests, useful though they may be, are a test for major design errors (and they have to be big), brittle fracture and leaks, but not full fitness-for-purpose in service. They do not address the effects of corrosion, cyclic stress, dynamic loads, process upsets, erosion, embrittlement or the many other damage mechanism that cause vessels to fail. Watch out for questions like this.

13.3 Final word: API exam questions and the three principles of whatever (the universal conundrum of randomness versus balance)

As with most engineering laws and axioms (pretend laws) you won't get far without a handful of principles (of whatever).

The first principle (of whatever) is that, faced with the dilemma between randomness and balance, **any set of exam questions is destined to end up with a bit of both**. A core of balance (good for the soul and technical reputation of the whole affair) will inevitably be surrounded by a shroud of some randomness, to pacify the technically curious, surprise the complacent and frustrate the intolerant, in more or less equal measure. There is nothing wrong with this; the purpose of any exam programme must be to weed out those candidates who are not good enough to pass.

Now we have started, the first principle spawns, in true Newtonian fashion, the second principle; a strategy for dealing with the self-created problems of the first. The problem of course is the age-old one of *high complexity*. Code documents contain tens of thousands of technical facts, each multifaceted, and together capable of being assembled into an almost infinite set of exam questions. We need some way to deal with this. **The second principle** becomes:

Selectivity can handle this complexity

Tightening this down, we get **the third principle:**

Only selectivity can handle this complexity

There is nothing academic about the third principle of whatever; it just says that if you try to memorise and regurgitate, brightly coloured parrot-fashion, **all** the content of any exam syllabus, you are almost guaranteed to fail. You will fail because most of the time the high complexity will get you. It has to, because exam questions can replicate and mutate in almost infinite variety, whereas you cannot. You may be lucky (who doesn't need a bit of luck?), but a more probable outcome is that you will be left taking the exam multiple times. Round and round and round you will go at your own expense, clawing at the pass/fail interface.

A quick revisit of the first principle of whatever suggests that being selective in the parts of an exam syllabus you study carries with it a certain risk. The price for being selective is that you may be wrong. Most of the risk has its roots in the amount of balance versus randomness that exists in the exam set. The more balanced it is, the more predictable it will be, and the better your chances. Do not misread the situation, though, your chances will never be any worse than they would have been if you had not been selective; the third principle tells us that.

Remembering this, you should only read the remainder of this book if you subscribe to the three principles *and* you think selectivity is for you. If you don't recognise the code references, clause numbers or abbreviations, then you need to start again at the beginning of the book, and get yourself a bit more background knowledge.

FIG 13.2
Exam question alert

– The character of API (SIFE) exam questions –
Here are five specific insights about API SIFE exam questions.
- They address **general inspection principles**, rather than specific engineering knowledge.
- There are **no calculation-type questions**.
- The BoK codes contain lots of **diagrams, figures and graphs**. Realistically, it is difficult to expect candidates to memorise detailed information from graphs for closed-book exam questions so questions based on these are typically quite general, which is good news.
- **Questions based on detailed reasoning and analysis** do not flow naturally from the content of API SIFE codes. They are used, but you can expect the reasoning required to be at a fairly low level. Simple choices such as the elimination of obviously incorrect options are more common than artificially constructed scenarios requiring long strings of logical thought.
- **Verbatim quotations of terminology, definitions, principles and concepts** fit best with API SIFE closed-book exam-type questions. They are stated in fairly tight, abbreviated form in the SIFE study document, rather than being elaborated by rigorous technical explanations. This is good news for the attention span of question setters, reviewers and approvers.

Chapter 14

The SIFE body of knowledge (BoK) and study guide

14.1 The SIFE BoK: what's in it?

As with all the other API examination programmes, the SIFE programme comes with its own body of knowledge (BoK). From API's viewpoint this represents the body of knowledge that source inspectors (SIs) of fixed equipment 'should know'. BoKs always represent brave choices – there is an almost limitless list of fixed equipment that an SI may be called to inspect, but for exam purposes, the line has to be drawn somewhere as to what is included and what is not.

Figure 14.1 shows the wide extent of the BoK. Added together, all these documents amount to several thousand pages of closely defined technical data. This is not a new approach; most API exam programmes work in the same way – a vast BoK from which practicality dictates a relatively small number of exam questions have to be extracted. For the SIFE programme, the examination comprises 100 questions, all of which have to be answered *closed book*. On the face of things, both choosing the questions (and answering them) looks an insurmountable task.

14.2 Dealing with such a large BoK (the road map analogy)

This huge BoK only looks huge to you, not to the people who write or agree to it. This is because their viewpoint influences them to see all the multiple codes, their figures, lists and tables of data as analogous to the multiple towns and villages shown on the *road map of a country*. Any country will have hundreds of thousands of these, many more than the

FIG 14.1
API SIFE
Exam question distribution

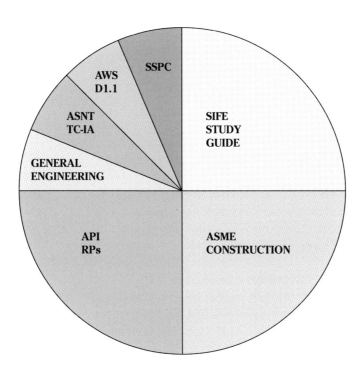

number of pages in the SIFE BoK. You wouldn't worry about using a road map containing all these data, so why would you be concerned about this (now relatively small-looking) BoK?

Are you now worried about how you would possibly be expected to know how many buildings are in village number 32 394 of the 100 000 shown on a typical country's road map. You've never been there, so how would you know that, if you got an exam question on it? Even if you had been there, and for some bizarre reason knew the number of buildings, this knowledge would be of no use to you if the question was in fact not about that village but village 32 395, a short distance down the road. Once again, the problem looks insurmountable.

Map reading

Before you embark on a whistle-stop building-counting tour of the 100 000 towns and villages on your map, think what would happen if the exam questions *were* about the number of buildings, or even how big they are, or what is contained in them. Forced into choosing 100 closed-book questions (meaning you can't go there to check) from an almost infinite set of town- and village-based data, then from say a set of 100 exam candidates, each candidate has only a 1 in 1000 chance of even *living* in the destination chosen, never mind knowing how many daffodils are in the flower pot outside number 23 High Street. On balance, all the candidates would get all of the questions wrong, all of the time, apart from any marks that could be picked up by a process of random guesswork.

This is good news because now you know that the questions *cannot be about the destinations* on the map.

So what are the questions about?

They have to be about the *routes that lead to* the destinations. Not only that, we know they cannot be about the minor roads, lanes, tracks and footpaths that lead to each, otherwise we would run into the same complexity problems that we saw with the data set of the towns and villages. The questions have to be about the *major routes* that lead to the important areas – the large cities and conurbations, plus other popular destinations such as holiday resorts, national parks and suchlike. These routes are easy to spot on our roadmap. They show as thick and brightly coloured, and it is not difficult to remember which ones go north or west. To make things easier, they are connected as a network, so large routes connect to other large routes. This is a now relatively easy mental picture to retain. No longer are we worried about the 100 000 destinations.

From routes to SIFE exam questions

Now let's apply this to the world of the API SIFE exam question. Despite the multitude of destinations (code pages), the majority of the questions *have to be* about the common directions that the codes take, rather than the content of all the pages themselves. To help you along, the programme managers have chosen to make all the questions *closed book*, fitting our model beautifully. If some questions had to be answered 'open book' then the game changes, and we would need to rethink the road map analogy.

You don't even need the map to know what the routes are that spawn the main body of SIFE exam questions. There are seven of them.

1 The language and *terminology* of source inspection (SI).
2 Which *parties* are involved and what do they do.
3 How does *quality surveillance* work?
4 What exactly is the *role* of the SI (you)?
5 Important *principles* of construction codes.
6 The use of cross-referenced *industry standards*.
7 Some *basic* technical knowledge about manufacture and testing of fixed equipment.

On balance these subjects form the basis for most of the SIFE exam questions. You can expect some of them to be dressed up in a bit of technical detail (particularly points 5 and 6), but the core content will be based around one (or more) of these core 'routes'.

Looking at this you can see how this neatly side-steps the problem of the huge complexity of all the different codes and pages listed in the BoK. It promotes the tendency for exam questions to be *general* rather than code-specific. Additionally, questions will address a subject that is referenced by *several parts* of the BoK: welder qualification, material traceability, or pressure testing, for example. This gives them extra validity (they think).

Looking briefly at each in turn, the titles of 'route points' 1–7 suggest the sources of questions that you can expect to have to answer. Figure 14.2 shows the details.

14.3 The SIFE study guide book

To complement the codes listed in the BoK, API have produced a 50-page document entitled *Guide for Source Inspection and Quality Surveillance of Fixed Equipment* (see Figure 14.3). This can be downloaded from the API website. In addition to acting as part of the exam BoK, it provides a useful and realistic background to the role of source inspection. In time, this may be issued as a more formal API recommended practice (RP) document. Note a few important points about this SIFE guide.

- It provides a useful elaboration on the code 'effectivity list' that makes up the BoK.

FIG 14.2
The seven 'routes' to API SIFE exam questions

1. **Language and terminology**
   - Abbreviation and acronyms
   - Documentation names
   - Code stamp meanings

2. **The parties**
   - Who is involved?
   - Who is responsible for what?
   - Where do responsibilities start and end?

3. **Quality surveillance**
   - What constitutes a quality management programme?
   - The role of the ITP
   - Who signs what?

4. **The role of the SI (you)**
   - Who does the SI interact with and how?
   - Verifying, checking, surveillance and compliance activities
   - Validating training records

5. **Principles of construction codes**
   - Material control
   - Fabricating material
   - Inspection and test practices

6. **Cross-referenced industry standards**
   - Common cross-code standards and recommended practice (RP) documents
   - ASME, API, AW
   - S and SSPC

7. **Basic technical knowledge**
   - Material types
   - Material joins (forging, castings, plates and pipes)

→ SIFE EXAM QUESTIONS

FIG 14.3
The SIFE study guide

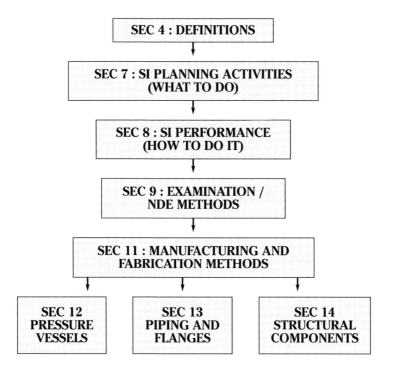

- It identifies specific technical points within referenced codes that are deemed important to SIs.
- *Large numbers of SIFE exam questions* are sourced from this document, in preference to the detailed content of the referenced codes. Its style and content make it suitable for API-style closed-book questions.

This third point is the most important, making the SIFE guide the most likely source of many of the examination questions. You can read it as it stands, and take in some useful information – the most effective way, however, is to look at a few of the most important sections, and awkward points, then attempt some sample questions.

SIFE guide section 4: Definitions abbreviation, acronyms

All API documents contain a definitions section, to clarify the meaning of terms and acronyms used in the document. Most are common sense and match the generally used meaning of the words, but a few have their own meaning. Key acronyms from section 4 are the following (see Figure 14.4).

- **EPC**: This is the engineering design construction company responsible for building a large plant. On completion, they hand it over to the owner/user.
- **S/V**: This represents the supplier/vendor. Typically they construct smaller plant assemblies (skid-mounted equipment for example) for supply to the EPC.
- **M/F:** The manufacturer/fabricator(s) who actually manufacture the component parts (vessels, pipework, valves and so on). There will be hundreds of these in a large EPC construction project.

Figure 14.4 shows the hierarchy of how these parties fit together. Next consider the other parties with involvement in the source inspection activity (see Figure 14.5).

- **The inspection co-ordinator**: works for the source inspection company, co-ordinating the inspection visits of SIs. The role also provides a link back to the function of administering inspection reports, NCRs and engineers' approvals of concessions to specification compliance.
- **SI**: The source inspector themselves (you). This title is not universal – in many industries and countries it is more commonly known as the *shop* inspector or *works* inspector.

Indications, imperfections and defects

Strangely, you won't find these technical terms listed in the SIFE Guide section 4. They pop up in many parts of the BoK and are part of a small family of terms that API pre-suppose you are familiar with already. Fortunately they are used with consistent meaning across all API codes. They are

- indication

FIG 14.4
Source inspection and quality surveillance
– parties' responsibilities –

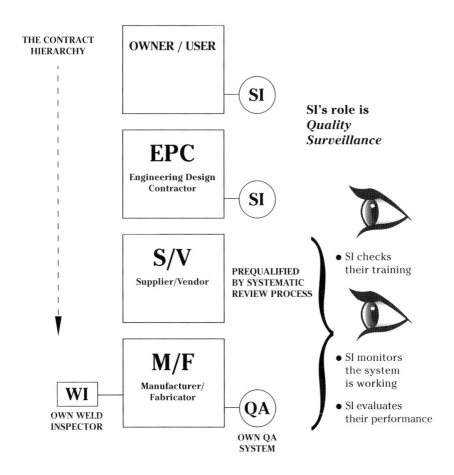

- imperfection, flaw or discontinuity (assume they all mean the same)
- defect.

If you don't know the difference, have a look now at Figure 14.6. Note their relevance to the results of visual examination or NDE. When an *indication* is found and determined to be a real physical finding on the

FIG 14.5
Inspection co-ordinator role

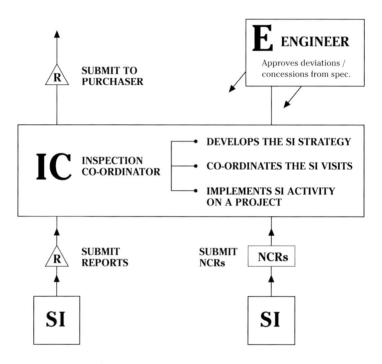

component (rather than an RT film mark, for example) then it becomes an *imperfection, flaw* or *discontinuity*. If it is then checked against code acceptance criteria and found to *exceed* acceptable limits, it is now classed as a *defect*. If not, then it remains an imperfection, flaw or discontinuity.

Finally, the MTR

The MTR is the *material test report*. Normally issued from the source steel mill or foundry, it shows the results of chemical analyses and mechanical tests, as well as batch numbers, heat numbers and other

FIG 14.6
Key definitions

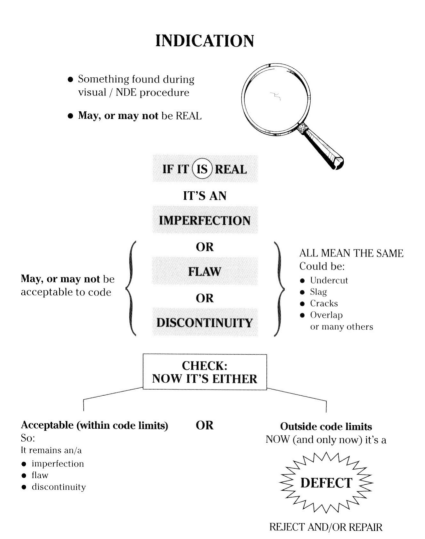

identifiers. The MTR is therefore evidence of *what the material is*. Alternative terms for the MTR would be

- mill certificate
- material certificate.

Be careful not to confuse the MTR with an MDR. The MDR (manufacturer's data report) is a specific requirement of the ASME code issued when a vessel is signed off as code compliant by the ASME authorised inspector (AI) qualified by the USA National Board. It is also known as form U-1 (see an example in Appendix 3).

SIFE guide sections 5, 6, 7

These are set out as

- section 5: *Training and certification*
- section 6: *Source inspection management programme*
- section 7: *Project specific source inspection planning activities.*

Section 5 *Training and certification* looks like an experimental part of the document. It breaks down the level of training and experience for SIs into four levels; entry level, basic level, stand-alone level and master level. The idea is consistent within itself, but is, as yet, not used in industry practice. It also does not fit in with the API SIFE qualification itself. For practical reasons this section of the study guide is not included in the SIFE BoK, so you can safely ignore it.

Section 6 *Source inspection management programme* makes good sense but offers less information than its title suggests. It is simply a management system, the purpose of which is to provide the SI with all the information they require to work on projects. The system is generic so can be used for all SI projects. The full list of 'outputs' from the management system is set out in bullet lists in section 6, and these are self-explanatory. In practice, you will find companies trying to manage SI activity *without* some of the necessary points in the section 6 lists, so the lists can provide useful real-world guidance. It is difficult to extract sensible closed-book exam questions from this section of the SIFE study guide, so do not expect to see more than the odd one appear in the exam.

Section 7 *Project specific source inspection planning activities* is a more specific version of section 6. The principle is given below.

- Source inspection should be directed towards the materials and equipment that *need it most.*

You can think of this as a risk-based approach to source inspection and, as with all risk-based assessments, it needs input from several parties and people to ensure it is done well. In real-world situations, not all projects get this right, leading to

- SI activity being wasted in multiple inspections on fairly dependable (low-risk) manufacturing processes and components
- high-risk, potentially more awkward, situations only being subjected to light surveillance or none at all, with inspection input limited to review of paperwork.

This is a classic scenario, explaining how serious non-compliances can slip through the SI activity that is supposed to catch them. Real risk is ignored, to be replaced by activities based on convenience or cheapness, NCRs are considered an embarrassment rather than a success, and everyone is happy. Self-congratulation abounds, while the real non-compliances slip through, only to make their appearance on the construction site, or during plant operation later on.

Section 7 partially addresses this scenario by conceding that equipment assessed as low risk and therefore not subject to much source inspection relies on the supplier/vendors' (S/V) internal quality system to identify and rectify non-compliance at source. In reality, sometimes this works and sometimes it does not.

The API view on 'high risk'

Section 7 gives a list of factors that API believe constitute manufacturing scenarios that are considered high risk. Figures 14.7 and 14.8 summarise the situation. Also shown in this figure are a couple that maybe *should have* been explicitly included. These are to do with so-called *rogue materials*.

- Material *origin*. There are unfortunately some material sources in some countries of the world where poor-quality materials are commonplace.
- Material *types*. To compound the problem above, some materials, because of their sensitivity to alloying elements or heat treatment, are just naturally *much more variable* in quality than others. With these, then, *quality* and *specification compliance* may not be exactly the same thing. It is easily possible to achieve specification compliance with a poor material if the specification is loose to start with. The mechanical properties of toughness (impact strength) and high-temperature creep resistance are generally the main issues here.

FIG 14.7
High-risk manufacturing scenarios

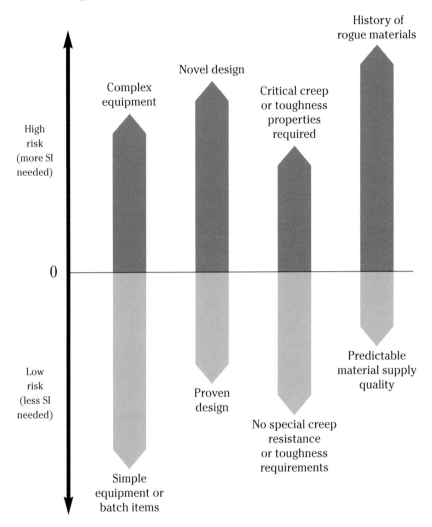

Read this in conjunction with the 'equipment risk assessment' list in the API SIFE study guide

FIG 14.8
Rogue materials risk factors
(Ref SIFE 6.2)

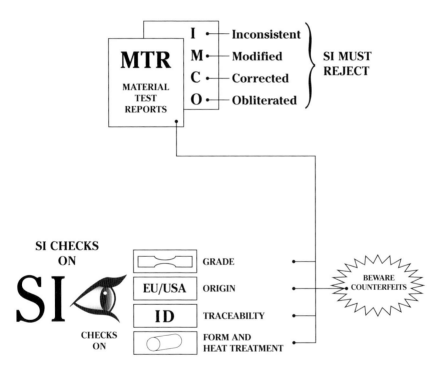

What is the output from the SI planning activity?

The output from this (it is a type of risk assessment, remember) is the inspection and test plan (ITP). Key points about ITPs raised in section 7 are as follows.

- Each type of equipment needs its specific ITP. Generic ones do not work well.
- ITPs should include reference to *acceptance criteria*.
- The inspection co-ordinator plays a large part in deciding the ITP activities derived from the risk assessment.

Implementing the ITP is *the* most important role of the SI. A well-set-out and detailed ITP brings an organised structure to the manufacturing inspection process, providing a focus for all parties involved. The SIFE study guide does not show a typical ITP, so you can see one in Chapter 8 of this book.

SIFE guide section 8: Source inspection programme

You can think of section 8 as a collection of explanations as to how the source inspection *process* is done, rather than containing technical details, as such. The list below shows the breakdown. We shall look mainly at sections 8.2 and 8.4 – taken together these contain more important information than the others.

8.1: *Inspector conduct and safety*
8.2: *Review of project documents*
8.3: *Performing the source inspection*
8.4: *Scheduled planning events (i.e. meetings)*
8.5: *Report writing*
8.6: *Non-conformances/deviations*
8.7: *Project continuous improvement*
8.8: *Source inspection continuous improvement*

Section 8.2: *Review of project documents*
This section is useful in showing the technical complexity involved in the role of the SI. It lists the scope of documents that can be referenced in an ITP, and which the SI can be called to check against. Figure 14.9 shows this scope in diagrammatic form. For large numbers of fabricated equipment items the total scope can amount to thousands of pages of directly relevant criteria, and many thousands of others containing general requirements and more peripheral stuff. Given this amount of material, you can see how familiarity with these documents has to be preceded by *selectivity* in deciding which bits are important and which are not.

Be careful of the contract hierarchy.

A quick review of section 8.2 of the SIFE study guide will show you that it does not mention the *contract hierarchy*, which allocates priority to the project document groups shown in Figure 14.9. All construction project contracts have a document hierarchy written into them, setting out which documents take priority over others. Normally the client's specification and in-house standards take priority over others, with general industry standards (AWS, SSPC etc.) residing near the bottom

The SIFE body of knowledge (BoK) and study guide 221

FIG 14.9
Project documents relevant to the work of the SI (you)

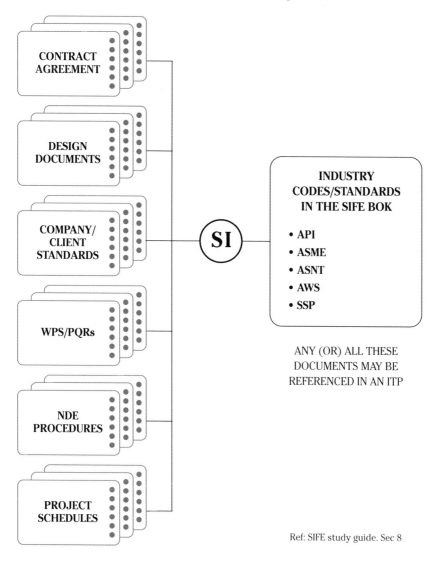

Ref: SIFE study guide. Sec 8

of the list. Paradoxically ITPs (a fundamental part of the quality surveillance process) often also lie near the bottom, wrapped up in the category of 'working documents'.

As an SI, expect to meet a few situations when there is a conflict of technical requirements between the content of documents that inhabit different levels in the contract hierarchy. It should rarely become a major issue, however, and occurs more by error than by design.

Shall and *should*

The words *shall* and *should* appear in multiple codes, contracts and specification documents, and have great significance in the world of the SI. Amazingly, there is (almost) universal agreement among English language codes about the contractual meaning of the words. The SIFE study guide definition therefore concurs with the API, ASME and other codes that make up the SIFE BoK. The rules are:

- **Shall** means *must, mandatory* or *without exception*. Non-compliance with a requirement containing *shall* therefore requires a non-conformance report (NCR).
- **Should** has a softer meaning; it is an expression of an *expectation* or a *good practice*. Its meaning can therefore vary with interpretation or viewpoint – in fact there may be several different correct versions of what it means. Non-compliance with a *should* clause does not warrant an NCR and is generally dealt with using a supplier's observation report (SOR).

On a practical note, as an SI you should put most of your investigative energy into finding NCRs. SORs are easier to initiate but can either fill the surveillance programme with trivia or grow into monsters, sucking up the technical enthusiasm of the parties into interesting technical discussions of little real consequence. NCRs beat SORs, every time.

Section 8.4: Scheduled planning events

Planning events are SIFE-speak for *meetings*. Section 8.4 identifies two meetings that form part of a source inspection programme: the pre-purchase meeting and the pre-inspection meeting. Of these, the pre-purchase meeting does not always happen, and is of little interest to the SI anyway. The pre-inspection meeting, however, generally does happen and ideally should be attended by the SI. On construction projects widely sourced across multiple contractors and countries, attendance is frequently pushed higher up the pecking order to involve co-ordinators, QA and supply chain managers, and all their friends.

Does this pre-inspection meeting have any output? It will depend on the state of the ITP when the meeting is held. Look at these two important points.

- To be successful, an SI *needs* a well-structured and detailed ITP.
- Good ITPs are predominantly *technical* rather than administrative. Having clear standards and acceptance criteria are more important to you than knowing who approved the ITP and who approved the approver.

Stripped bare, the only real output of a pre-inspection meeting should be to finalise the technical content of an ITP. As there are many of them, to fit in with hierarchy of contracts and subcontractors, it becomes very difficult to cover all the various versions at one go. This can be bad news for *technical completeness* of ITPs. Suddenly, it becomes easier for discussions to drift towards the territory of easy administrative issues, such as who signs off ITPs, or who receives copies, rather than to dig into awkward discussions about detailed technical issues. The higher the hierarchical level of the attendees, the less they will understand the technical issues anyway, so they will be deferred to another day. Now you can see why source inspection can be tricky; you may be left to debate and discuss during the source inspection itself what some of the specific technical requirements actually *are*.

This does not always happen of course – some ITPs emerge complete and comprehensive for you to work to. Code references and clear acceptance criteria are then easy to check and NCRs easy to decide when you have the power of a strong ITP behind you. Others are poor and spineless, leaving you on your own. Which one you get depends on the contract parties and their structures. Country and culture also play their part.

Remember the summary points about these meetings.

- The *pre-inspection meeting* is the important one.
- The SIs *may* or *may not* attend (that's what the SIFE study guide says).
- If the ITP is *good* it will help you immensely, diffusing technical arguments and differences in interpretation about your NCRs.
- If the ITP is *weak*, you as an SI will be in a weakened position. Your NCRs will be more difficult to issue and prove easier to overturn.

Section 9: Examination methods, tools and equipment

This section is more about source inspection *activities* rather than tools and equipment. It is broken down into the following.

9.2: *Materials of construction*
9.3: *Dimensional inspections*
9.4: *Visual inspection*
9.5: *NDE techniques*
9.6: *Destructive testing*
9.7: *Pressure/leak testing*
9.8: *Performance/functional testing*
9.9: *Surface preparation/coating inspection*

Most of these sections comprise just a brief summary of the activity, supplemented by cross-references, explicit or general, to codes that cover the activity in detail. SIFE exam questions are therefore general in nature and based on basic knowledge about fixed equipment and the source inspection process. For more code-specific detail (the exam is all closed book remember) then the best place to start is with the information in Part A of this book. Most of what you need is covered there – you can then reinforce this by looking at the individual code clauses referenced in the SIFE study guide section 9.

Section 9.2: Materials of construction

This is worth looking at separately as it is the most useful part of section 9. It is an obvious source of SIFE exam questions, not least because it has strong practical application. Rogue materials are an increasing problem, and it is the SI's job to find them. All the points listed in section 9.2 make sense and translate well into exam questions. API 578: *Material verification programmes* is in the SIFE BoK. It covers positive material identification (PMI) tests and techniques; note, however, that it is concerned more with preventing mix-ups between carbon steels and alloy steels than the issue of counterfeit materials from unreliable sources.

The other sections of the SIFE study guide

The remaining sections (10–14) of the SIFE study guide are made up of brief summaries of relevant bits of the following codes.

- Metal forms, properties and metallurgy (from API 577)
- Weldability and hardenability of metals (also from API 577 but more detailed)

- Basic information on pressure vessels (from API 572)
- Basic information on piping and valves (from ASME B31.3, B16.5 and API 598)

We will cover these in subsequent chapters. Again, you can use the study guide as a lead-in to reading the referenced code sections themselves.

The ASME and National Board (NB) code stamps

ASME and NB code stamps are commonplace in the USA and other countries that follow US practice. They are mainly found in the oil and gas industry. The importance of these code stamps is that they signify that a manufactured item of pressure equipment has successfully complied with *all* the requirements of the applicable ASME code(s) to the letter, including all documentation and certification requirements. This differentiates it from the large amount of equipment manufactured around the world that is constructed to 'ASME intent'. Such equipment complies with some parts (i.e. the *intent*) of the ASME code but not *all* of it. Frequently these are issues that have no real effect on the integrity or functionality of the item, that is, they are more of an administrative non-compliance than an engineering one. Equally, engineering design, materials or fabrication issues may just be non-compliant on a code-wording technicality, rather than anything that is technically important.

For fully code-stamped ASME equipment, the stamps used are reproduced in Annex A of the SIFE study guide (Figure 14.10 shows these). Note the change of stamp format that was introduced in 2013. The size of the stamp letters changed and the strange 'reversed P' letter was removed from the pressure piping stamp. Existing pre-2013 equipment (and its certification) will still carry the old versions. These will only be relevant for repair of existing equipment rather than source inspection of new constructions.

14.4 SIFE exam preparation

Figure 14.3 showed a broad outline of what is included in the SIFE study guide. Have a look at the full list of contents of the study guide and relate it to Figure 14.3. Note how section 5 *Training and certification* features in the contents list, but not in the figure; this is because it is not really an examinable subject for the SIFE exam. Section 5 is probably there for use if (or when) the study guide is adopted as a

FIG 14.10
ASME code stamps: boilers

ASME BPVC Section I:
Power boilers

- [S] Power boilers
- [A] Power boiler assemblies
- [E] Electric boilers
- [M] Miniature boilers
- [P] Pressure piping
- [V] Power boiler safety valves

ASME BPVC Section VIII:
Pressure vessels

- [U] Pressure vessels
- [UM] Miniature vessels
- [UV] Pressure vessel safety valves
- [UD] Pressure vessel rupture discs
- [U2] Alternative rules for pressure vessels

ASME BPVC Section VIII:
Pressure vessels

- [U3] High-pressure vessels
- [UV3] Safety valves for high-pressure vessels
- [RP] Fibre-reinforced plastic pressure vessels

ASME BPVC Section IV:
Heating boilers

- [H] Cast iron heating boilers
- [H] Heating boilers, other
- [HLW] Lined potable water heaters
- [UV] Heating boiler safety valves

ASME BPVC Section XII:
Transport tanks

- [T] Transport tanks
- [TV] Transport tanks safety valves
- [TD] Transport tanks pressure relief devices

Code stamps shown courtesy of American Society of Mechanical Engineers (ASME)

formal API code or recommended practice (RP) document in the future. For the moment don't worry too much about it.

The remainder of the SIFE study guide *is* of importance. You can expect perhaps 35–45% of the exam questions to be lifted directly from this guide – some using verbatim wording and others based on general principles, in association with some specific technical wording.

The six-step methodology

Our exam preparation recommendations all follow a similar pattern. You can see this pattern emerge in the treatment of this SIFE study guide. Following this methodology will give you the best chance of doing well in the API examination. The methodology is as follows.

1. Read the initial summaries provided in the following sections of this book.
2. Do the activities that each suggests, including any required code mark-ups (by tab, text or post-it note – it's up to you).
3. Answer the (fairly broad) questions that each poses. You should do this by a careful reading of the relevant documents, rather than just skating over the content.
4. Look at the **'ARE YOU SURE?'** box that follows the text. Once you can answer 'yes' to this box then you are ready to try the set of quiz questions that follow it.
5. Attempt the quiz questions. These are **open-book** question exercises so you should use the relevant documents to *find* (not guess) the answer to the questions. Once you have found the answer, then *mark your code copy* by underlining the answer you have found and writing 'Q' in the nearest margin. These will act as reminders for your revision.

YOUR TARGET

Your target should be to obtain a mark of 60% + at every first attempt.

If you get lower than this, you should conclude that you are not doing sufficient preliminary learning before attempting the quiz (revisit steps 2–4 above).

SIFE Section 4: initial summary

Most API codes and RP documents contain a section near the beginning covering *the definitions* of terms used in the remainder of the document. Sometimes these follow the conventional English language meaning of the words and sometimes they don't – their meaning may be adapted to suit better the context of the subject or the messages the particular document is trying to convey.

Section 4 of the SIFE study guide covers *definitions*
Read through this section and make sure you know the meaning of the names or abbreviations referring to the various parties involved in the

manufacturing and source inspection process (then mark them up in your code). The main ones are

- EPC
- examiner
- inspection agency
- inspection co-ordination
- M/F
- quality surveillance (and the people who do it)
- SDO
- SI
- SME
- S/V.

Not all of these are as obvious as you might expect, so make sure you know who they all are.

What do the various parties actually *do*?
In the world of API examinations, great importance is placed on the issues of **who does what** in the manufacturing and inspection process. Another way of describing this is *roles and responsibilities*. In some cases the practical aspects of an activity will themselves control who is best suited to approve it or authorise it to start. In other cases published codes and documents express a preference or an instruction on who should approve or authorise an activity. API documents work like this.

NEXT STEP: Have a scan through the entire SIFE study guide document identifying where (and in which sections) these roles and responsibility controls appear. Mark up the main ones that you find (with the 'Q' mark in the margin remember). Pay particular attention to the roles and responsibilities allocated to the SI; that's you.

ARE YOU SURE?
That you have identified and marked up sufficient clauses that explain the role of the source inspector (SI)?

If you are, then attempt question set 14a at the end of this chapter.

Now turn to the SIFE study guide section 8

Section 8 is a long section of the SIFE study guide covering ten or so pages entitled *Source inspector performance*. It is a mixture of guidance on the roles and responsibilities of source inspection at the beginning and end of the section, interposed with a BoK-based list of relevant codes and reference documents in the middle.

Have a look through and identify and mark up those entries relating to the following.

- Which *documents* the SI is required to review.
- The actual activities of *performing the source inspection* (section 8.3). There is a lot of useful information in here.
- The difference between the *pre-purchase* meeting and *pre-inspection* meetings, with their views on who should be doing what.
- Finally, the topic of *continuous improvement* is an almost guaranteed subject for SIFE examination questions. It oozes possibilities for questions related to general principles and verbatim wording. Make sure you understand thoroughly the description of it in sections 8.7 and 8.8.

ARE YOU SURE?
That you have a clear view of what the SI is responsible for?

And could you also answer questions on what the SI is **not** responsible for?

If you are, then attempt question set 14b at the end of this chapter.

Finally, some selected points from SIFE study guide sections 9, 12 and 13

Section 9 is a generic section of the SIFE study guide that summarises the various NDE and other tests that make up the 'testing' activities of source inspection. Have a look through this section, noting how it contains not just the NDE activities with which you are probably familiar, but also visual and dimensional testing (sections 9.3, 9.4) and pressure/leak testing activities in the later section 9.7.

Hidden away in section 9 are five other statements on the roles and responsibilities of various parties, including those of the SI.

Now do this task: go through section 9, identify the points that define what any of the parties *actually do* or *are responsible for*. Mark these up clearly, as you did for the earlier sections.

SIFE study guide sections 12 and 13 differ from the other sections in that they relate to specific types of mechanical equipment: section 12 (vessels) and section 13 (piping and valves). Once again you need to identify the few rules and responsibilities requirements hidden away in these sections, marking them up in the same way.

An important point: don't confuse the SI and AI

Remember that for ASME VIII vessels built in the USA and other countries accepting ASME jurisdiction, the ASME BPVC requires the presence of an authorised inspector (AI). Do not confuse this with the role of the API source inspector (SI). Just think of the SI as a more general, less ASME-specific incarnation of the AI, not restricted to the code requirements of the ASME BPVC, and with a wider, less code-prescribed role and duties.

ARE YOU SURE?
That you have a good understanding of the SIFE guide sections 9, 12 and 13 (and have also looked at the content of section 11)?

If so, attempt question set 14c at the end of this chapter.

TO COMPLETE THIS PART OF YOUR PREPARATION
PLEASE DO A SERIOUS REVIEW OF THE SCORES YOU ACHIEVED IN THE QUESTION SETS 14a, 14b AND 14c. IF YOU PASSED WITH 6/10+ IN ALL THREE SETS...WELL DONE

If you scored less than 6/10 for any of them, you may have misunderstood the level of effort and concentration required to pass the API SIFE exam.

SO CHECK WHAT YOU DID
DID YOU?

- Do the necessary reading and marking up as requested?

AND

- Did you look up the answers to the questions in your codes as instructed? Or did you just guess them?

If you ignored these instructions, then this is the main reason for your poor marks.

Question set 14a: The source inspection process and responsibilities

Q1. Purpose of the SI programme

The API source inspection programme is:

(a) Applied to all suppliers/vendors (S/Vs)
(b) Only applied to S/Vs that have a quality management process in place
(c) More applicable to S/Vs that do not have a quality management process in place
(d) To assist S/Vs in ensuring inspection and test plans (ITPs) are working as they should

Q2. Purpose of SI programme

An SI acts to determine:

(a) Whether purchased equipment meets contractual (paperwork) terms and conditions
(b) Whether the ITP is correct
(c) Whether purchased equipment is 'fit for purpose'
(d) All of the above

Q3. SI programme responsibilities

The responsibility for verifying that an S/V's quality management process is working as it should belongs to:

(a) The S/V's QA certification/accreditation body
(b) The S/V's management themselves
(c) Any SI appointed by the plant EPC or owner/user
(d) The purchaser

Q4. Guidebook definition

Confirmation that a person meets the requirements for a specific qualification is called:

(a) Competence
(b) Qualification
(c) Certification
(d) Accreditation

Q5. Guidance definition: inspection agency

If a manufacturer (S/V) has a fully documented and certified QA/QC system, then it cannot act as:

(a) An inspection agency
(b) An S/V for non-ASME equipment
(c) An EPC
(d) A designer of ASME equipment

Q6. Guidebook definition: responsibility of inspection co-ordinator

The individual responsible for the implementation of source inspection activities on a project is:

(a) The person who signed off the ITP
(b) The inspection co-ordinator
(c) The source inspector themselves
(d) The source inspector's manager

Q7. SI guidebook definition

An NCR is a report filled out by:

(a) The examiner
(b) The S/V's QA representative
(c) The SI
(d) The inspection co-ordinator

Q8. Guidebook definition

Source inspection means the same as:

(a) Quality assurance
(b) Quality management
(c) Quality control
(d) Quality surveillance

Q9. Guidebook definition

Source inspection management programmes are:

(a) Developed by the inspector co-ordinator
(b) Generic in nature
(c) Agreed with the purchaser/client

(d) Part of the inspection planning activities

Q10. Inspection planning

A project-specific inspection plan should be developed by:

(a) The QA engineer
(b) Project planning personnel
(c) The source inspector
(d) The inspection co-ordinator

Question set 14b: The source inspection process and responsibilities

Q1. Project document review

Prior to commencing quality surveillance specified in the ITP, the SI should check that the S/V has the most current documents specified in the engineering design and:

(a) The ITP has been signed by the manufacturer
(b) The equipment is ready for inspection
(c) All project documents have been approved by the purchaser
(d) QA accreditation certificates are available for review

Q2. NDE responsibilities

Who is responsible for developing NDE procedure to be used on a specific project?

(a) The purchaser and owner/user
(b) The purchaser and S/V
(c) Examiner (NDE-qualified)
(d) S/V

Q3. Project schedule responsibilities

The SI is generally not responsible for:

(a) Slippage of milestone progress
(b) Monitoring delivery of project-specific equipment
(c) Reporting on fabrication status
(d) Verifying evidence that S/V fabrication personnel are properly trained

Q4. Source inspection work process changes

If an S/V proposes work process changes that could impact cost, schedule or quality then:

(a) The SI should review and approve them if applicable
(b) The SI should inform the inspector co-ordinator to plan them into the schedule
(c) The SI should request them in writing to pass to QA/expeditor staff
(d) They must be submitted for review to the purchaser

Q5. SI responsibilities for pre-purchase meeting

The SI:

(a) Does not have to participate in the pre-purchase meeting
(b) Shall participate in the pre-inspection meeting
(c) Shall participate in the pre-purchase meeting
(d) Shall participate in both pre-purchase and pre-production meeting

Q6. Inspection responsibility

SI reports should first be submitted to:

(a) The manufacturers (S/V)
(b) The purchaser
(c) The inspection co-ordinator
(d) All of the above together

Q7. NCR responsibility

Once a non-conformance has been identified the SI should inform, as soon as practical:

(a) The manufacturer's (S/V) quality representative
(b) The inspection co-ordinator
(c) The responsible engineer
(d) The purchaser's representation

Q8. NCR concession (disposition)

When is it the SI's responsibility to ensure that an 'NCR disposition' has been properly implemented?

(a) When it has been approved by the purchaser
(b) It is not normally the SI's role

(c) When it has been agreed by all parties
(d) Always

Q9. Deviation from specification: Responsibilities

During an inspection:

(a) The SI shall approve deviation items on the ITP
(b) Deviation from specification must be approved by the responsible engineer
(c) Deviation from specification shall be approved by the SI
(d) Deviation from specification shall be authorised by the SI and the purchaser

Q10. SI project continuous improvement

Source inspection continuous improvement recommendations identified at the completion of source inspection activities should include:

(a) Ways in which the S/V can improve their accredited QA system
(b) A schedule of NCRs and any subsequent disposition
(c) An evaluation of the performance of the S/V
(d) An evaluation of areas for cost reduction

Question set 14c: Source inspection process and responsibilities

Q1. Pressure test responsibilities

Pressure/leak testing during a source inspection is normally specified by:

(a) Codes/standards and contractual agreement
(b) Health and safety (H&S) personnel at the S/V site
(c) The ITP
(d) The SI in conjunction with S/V site H&S personnel

Q2. Pressure testing SI responsibilities

Before witnessing a pressure test at S/V premises it is the SI's responsibility to check:

(a) Whether a hydraulic or pneumatic test is the most practical and safest option
(b) That a PRV is fitted to prevent over-pressurisation

(c) That a correctly designed PRV is fitted and that the test has been witnessed by an authority
(d) That safety barriers are in place at least 60 feet (18.3 m) from the test location and that all personnel nearby have been informed

Q3. Manufacturer's responsibilities

Compliance with codes, standards and specifications contained in the contractual agreement is the responsibility of:

(a) The purchaser (on behalf of the owner/user)
(b) The manufacturer/fabricator (M/F)
(c) The source inspector (SI)
(d) All of the above parties

Q4. SI responsibilities at M/F

When performing source inspection at M/F premises, the SI is not responsible for:

(a) Identifying NCRs involving complex technical issues
(b) Following the ITP
(c) Surveillance of ITP-related documentation
(d) Compliance with codes and standards

Q5. M/F processes

During source inspection, rework and repair activities should be:

(a) Accurately documented in the ITP
(b) Authorised by the SI
(c) Approved by the purchaser
(d) Authorised by the purchaser

Q6. SI responsibilities during vessel dimensional check

When conducting dimensional checks of an ASME VIII-I pressure vessel, the SI should check for nozzle and attachment orientation, out of roundness for shell, weld mismatch and:

(a) Visual weld imperfections
(b) Joint efficiency and RT grade
(c) Mill under tolerance
(d) Correct location of tan line marking as specified in e.g. ASME VIII

Q7. SI responsibilities: Valve inspection

When performing source inspection on cast valves the SI should check against the purchase order for value size, material, rating and:

(a) Flow coefficient
(b) Weight
(c) Trim
(d) Wall thickness tolerance

Q8. General approach

Normally, which of these does an SI not have the responsibility (or power) to do when conducting a source inspection?

(a) Issue an NCR without the agreement of the manufacturer (M/F) or vendor (S/V)
(b) Issue an NCR before completing the inspection
(c) Stop manufacture when poor-quality work is found
(d) Make statements about subjects in which he/she is not an expert

Q9. SI-v-AI

When an ASME vessel is being manufactured under the jurisdiction of an ASME authorised inspector (AI), why is there a need for an additional SI to be involved?

(a) Because all AIs need to be monitored
(b) Because ASME codes do not cover all purchase order requirements
(c) It is often a requirement for legal and insurance reasons
(d) There is not – it just increases cost

Q10. SI -v- AI

When an ASME vessel is being manufactured under the jurisdiction of an ASME authorised inspector (AI) what title does ASME VIII allocate to the API SIFE inspector?

(a) None
(b) 'The inspector'
(c) 'The source inspector'
(d) The 'owner-user's representative'

Chapter 15

Metallurgy and materials: API RP 577 and related information

15.1 Introduction: metallurgy and materials – an important subject

Metallurgy and materials are important subjects for the source inspector. As we saw in Chapter 5 in Part A of this book, you need a good knowledge of the mechanical properties of materials if you want to be able to understand the content of material certificates included in new construction QA document packages. Of the code documents included in the API SIFE BoK, all (with the exception of the SSPC surface preparation codes) contain requirements for selecting, specifying and checking suitable materials of construction.

What level of knowledge is required?

You don't have to be a metallurgist to be a good source inspector (SI) (the two skills sets do not fit particularly well together anyway). You do need a basic appreciation of a few important parts of the subject, however. Have a look at Figure 15.1, which summarises the situation. Looking at this should certainly give you an idea of what you *can't do* if you lack the basic knowledge required.

Do you have this knowledge already?

Our experience shows that you probably do not – at least if you are from an inspection background. You will have more knowledge if you have worked in a material test house or have a qualification in metallurgy, but not all inspectors have done this. A common weakness is that source inspectors may not be fully confident with

- what all the values on a material certificate mean

FIG 15.1
SI knowledge: Metallurgy and materials

SIs NEED KNOWLEDGE OF:	TO BE ABLE TO:
MATERIAL TYPES • Carbon steels • Alloy steels • Non-ferrous alloys • Weld consumable materials	• Check the correct materials have been used in construction • Understand manufacturing and welding defects common to each
MATERIAL FORMS • Plates/pipes • Forging/wrought • Castings	• Anticipate problems common to these different forms
MATERIAL MECHANISM PROPERTIES • Strength • Hardness • Toughness • Ductility	Understand material certificates
MATERIAL TREATMENTS • Normalising • Tempering	Understand their effects on mechanical properties (and in-service damage mechanisms)

- typical errors or uncertainties in material properties that can cause a material to be unsuitable for the purpose for which it was intended
- materials knowledge in general.

15.2 The role of API RP 577 *Materials and metallurgy*

For the API (SIFE) examination most of the core materials knowledge that you need is provided for you in two documents

- the API SIFE study guide
- API RP 577.

Practically, most of the detail is contained in API RP 577. Have a look now at the index of API 577; it should look like the list below.

API RP 577 Contents

1. Scope
2. Normative references

3. Terms, definitions, acronyms
4. Welding inspection
5. Welding processes
6. Welding procedures
7. Welding materials
8. Welder qualifications
9. NDE
10. Metallurgy
11. Refinery and petrochemical plant welding issues

This is followed by Annexes A–E and 60+ figures and tables.

You can see that this is a fairly comprehensive list of subjects; there are a lot of points to cover. To fit in with this, this chapter consists of the following question sets.

Question set 15a: Material properties, strength and ductility
API RP 577 (10.3–10.4)

Question set 15b: Ductility, hardness and toughness
API RP 577 (10.4)

Question set 15c: Toughness, hardness and heat treatment
API RP 577 (10.4–10.6)

Each question set contains ten questions. Prior to starting any questions you should review the relevant content of API 577 (section 10) to give you an idea of how the subjects fit together. Then, before attempting each individual question set, read through the 'ARE YOU SURE?' instruction and review the code areas or subjects it recommends. You will need at least 30–60 min per set to do this properly.

Please remember that this preliminary learning is essential to prepare you for the questions. Do the questions 'open book' to help your learning (rather than your guessing).

First: Question set 15a: Material properties

Question set 15a covers the general subjects of physical and mechanical properties of metals. These are important because they feature in material certificates for new construction and repaired components.

Before attempting question set 15a

- Read through API RP 577 sections 10.3 and 10.4

until

- You understand fully the *difference* between physical properties and mechanical properties of a metal

then

- Have a look at a typical material certificate and identify where the values dealing with material strength and ductility are.

ARE YOU SURE?

- You understand the difference between yield strength and tensile strength and how they are linked to the property of *ductility*.
- You understand what *strain* is.

If you are, now attempt question set 15a

Next: Prepare for question set 15b: Ductility, hardness and toughness

Your learning resource for this subject is mainly contained in API RP 577 (sections 10.4.3 to 10.4.5). Read through these sections making sure you understand the definitions of

- *ductility* (it's to do with plastic deformation after the yield point)
- *hardness* (the only mechanical property of a metal that relates specifically to the surface of the material, rather than its full volumetric form)
- *toughness* (the opposite of brittleness).

When reading the API RP 577 sections, concentrate not only on the definitions of these three mechanical properties but also on

- how they are *measured* (i.e. what type of test is used)
- how the properties relate to each other. They all have interrelationships of some sort.

ARE YOU SURE?
That you understand the difference between ductility and toughness?

Tip: Think of ductility as *causing* toughness. If a material is ductile, then as a small crack or defect tries to propagate, the material will flow around the crack tip, discouraging it from propagating (this gives toughness).

Now attempt question set 15b

Next: Prepare for question set 15c: Toughness, hardness and heat treatment

This will revisit the mechanical properties of toughness and hardness but bring *heat treatment* into the equation. The subjects are covered in API RP 577 (sections 10.4–10.5). Read these again, marking the answers to questions such as the following.

- What is the relationship between hardness and resistance to sour (H_2S) service? (Why should they be related at all?)
- How is the test for toughness (the Charpy or Izod test) actually carried out? What temperature is it done at and how many test specimens are used?

Now a bit on heat treatment. Welding a metal adds large amounts of heat, causing significant effects to its microstructure. Steel, strangely enough, does not actually *like* being steel – its microstructure has been artificially created so it always maintains the potential to react unfavourably (crack) if further heat is added to it. To minimise problems with cracking, as-welded metal may need heat treatment before welding (preheat), after welding (PWHT) or both. This depends on thickness and the content of alloying elements such as carbon, chromium, nickel and a few others.

When reading API RP 577 (sections 10.4–10.5), concentrate on identifying the following.

- When is preheat required before welding?
- What is the governing thickness for preheat when two dissimilar thicknesses are to be joined?

ARE YOU SURE?
You really understand the *purpose* of preheat?
    If so, now attempt question set 15c

Your results?

The objective was for you to get at least 60% of each of the question sets on your first attempt. Please provide your own critical assessment of the scores that you achieved. Material properties are an important feature of all fixed equipment design codes. They are referred to in sections on strength calculation, welding, heat treatment and pressure testing. For this reason they feature regularly in SIFE exam questions.

Metallurgy and materials: API RP 577 and related information 243

Question set 15a: Material properties API 577

Q1. API 577 Physical properties of metal

A physical property of a metal or alloy is:

(a) Dependent on its metallurgical structure
(b) Insensitive to its metallurgical structure
(c) Measured using a physical test including force (e.g. tensile or impact)
(d) A property that is strength-related in some way

Q2. API 577 Physical properties of metal

The melting point of a metal alloy is a:

(a) Mechanical property
(b) Chemical property
(c) Physical property
(d) All of the above

Q3. API 577 Thermal expansion coefficient

When welding dissimilar metals (e.g. carbon steel to stainless steel) their different coefficient of thermal expansion can cause a sudden risk of failure in use due to:

(a) Galvanic corrosion
(b) Lack of weld penetration
(c) Distortion
(d) Thermal fatigue

Q4. API 577 Metal density

Castings and welds may contain imperfections such as isolated porosity and inclusions found using RT. These imperfections are:

(a) Less dense than the parent material
(b) More dense than the parent material
(c) Defects
(d) Crack-initiators

Q5. API 577 Mechanical properties of metals

Mechanical properties of metals of interest to source inspectors are yield strength, tensile strength, ductility and:

(a) Young's modulus
(b) Resistance to surface indentation
(c) Chemical composition (e.g. C, Cr, Ni etc.)
(d) Malleability

Q6. API 577 Yield/tensile strength

The stress on a material is defined as:

(a) The amount it elongates (stretches) under a given load
(b) The point at which it breaks
(c) The load divided by the cross-sectional area
(d) Its yield point

Q7. API 577 Stress/strain

A tensile test specimen shows the following dimension
Length before test = 100 mm
Length after breaking = 110 mm
Cross-sectional area before test = 90 mm^2
Cross-sectional area after breaking = 80 mm^2
From these data, the calculated value of the strain is:

(a) 1%
(b) 8%
(c) 10%
(d) 20%

Q8. API 577 Yield/tensile strength

For vessel or pipework pressure design purposes the main material mechanical property for consideration is:

(a) Yield strength
(b) Ultimate tensile strength
(c) Fracture strength and toughness
(d) All of the above

Q9. API 577 Ductility

Ductility is determined using a:

(a) Hardness test
(b) Charpy test
(c) Tensile test

(d) Bend test

Q10. API 577 Ductility

Ductility is the ability of a metallic material to:

(a) Stretch elastically
(b) Deform plastically
(c) Resist high stresses
(d) Avoid surface hardening in use

Question set 15b: Material properties API 577

Q1. API 577 Ductility

During a tensile test one of the measures of the ductility of the material is its:

(a) Yield point
(b) % reduction in area
(c) Breaking stress
(d) Original gauge length (e.g. 50 mm)

Q2. API 577 Ductility

Material that exceeds its yield strength in use may become:

(a) Softer
(b) Tougher
(c) Harder
(d) Weaker

Q3. API 577 Ductility weld bend test

The purpose of a bend test on a welded component is to check:

(a) For material compatibility
(b) Whether the weld will distort in use
(c) The tensile strength of the weld
(d) The ductility of the weld

Q4. API 577 Hardness

Hardness of a metal is its:

(a) Ability to deform elastically under stress in use

(b) Ability to resist fatigue stress
(c) Resistance to bending
(d) Resistance to surface indentation

Q5. API 577 Hardness

Hardness of a metal can be measured on scales of Brinell, Vickers, Rockwell or:

(a) Shore
(b) Knoop
(c) Newtons
(d) Charpy (Joules)

Q6. API 577 Hardness

In some low-carbon steels there is a (very) approximate relationship between hardness and:

(a) Thickness
(b) Impact strength (Charpy)
(c) Yield strength
(d) Tensile strength

Q7. API 577 Hardness

Which of these is not a designation used for hardness of a metal?

(a) HRC
(b) HV
(c) DPH
(d) BH

Q8. API 577 Hardness testing

Which of these material mechanical properties can be measured using portable test equipment?

(a) Hardness
(b) Toughness
(c) Tensile strength
(d) Ductility

Q9. API 577 Toughness

A low-carbon steel property of 70 Joules for a pipe or vessel steel indicates clearly it:

(a) Has low strength, low ductility
(b) Has high strength, low ductility
(c) Has high resistance to crack propagation under stress
(d) Has low resistance to crack propagation under stress

Q10. API 577 Joules

Many vessel or pipework codes require that Charpy impact (toughness) tests on the construction material be performed:

(a) After the hydrotest
(b) At CDTP
(c) At MAWP
(d) At MDMT

Question set 15c: Material properties API 577

Q1. API 577 Hardness-v-H_2S resistance

In some cases, the hardness of a material can affect its resistance to corrosion under sour (H_2S) service conditions. Resistance to wet H_2S cracking is improved if hardness is kept:

(a) Above 200 HB
(b) Below 100 HB
(c) Below 22 HRC
(d) Above 22 HRC

Q2. API 577 Toughness tests

Impact toughness (Charpy/Izod) tests are typically performed:

(a) With a single test specimen
(b) With at least 5 test specimens
(c) At various temperatures
(d) Using the same test pieces used for the tensile test

Q3. API 577 Test sample

Test pieces to check material toughness are typically:

(a) Flat 20 mm × 5 mm
(b) Square 10 mm × 10 mm
(c) Circular, diameter 12.5 mm ($\frac{1}{2}$ in)
(d) Tapered, to allow for varying cross-sections in use

Q4. API 577 Toughness testing

An impact strength test is also known as:

(a) A cantilever test
(b) A nick-break test
(c) A notch toughness test
(d) A slot test

Q5. API 577 Pre and PWHT

The purpose of preheating a low-carbon or alloy steel before welding is to:

(a) Help encourage the formation of martensite
(b) Increase the cooling rate to prevent hardening
(c) Decrease the cooling rate to encourage hardening
(d) Reduce the chances of HIC

Q6. API 577 Preheat

If preheat is specified in a weld WPS it should also be applied:

(a) As 'bakeout' after welding
(b) Before tack welding
(c) To all materials with the same P number
(d) To the consumables themselves between weld passes

Q7. API 577 Material testing

Destructive testing of steels is not usually performed to determine:

(a) Toughness
(b) Hardness
(c) Ductility
(d) Strength

Q8. API 577 Material testing

A Vickers or Rockwell test is not used to determine:

(a) Hardenability
(b) Hardness
(c) Resistance to surface indentation
(d) Resistance to ductile surface indentation

Q9. API 577 Preheat

As a general guide for a weld specified with a final weld bead width of 1 in, the total width of the preheat 'soak band' should be a minimum of:

(a) 2 in
(b) 3 in
(c) 4 in
(d) 5 in

Q10. API 577 Preheat governing thickness

A weld joint is joining:
   1 in thick material with a code-required preheat 300°F
   1½ in thick material with a code-required preheat of 400°F
What preheat temperature should be used for the joint?

(a) 300°F
(b) 350°F
(c) 375°F
(d) 400°F

Chapter 16

Non-destructive examination (NDE) – including ASME V and API 577

16.1 Introduction: the importance of NDE knowledge

Knowledge of NDE techniques is a fundamental requirement of being a successful source inspector (SI). Although some SIs may have obtained previous NDE qualifications, this is not actually essential. SIs do not usually perform NDE, but the role does include monitoring the techniques and results from those who do (i.e. the NDE 'examiners' and also 'interpreters' who may be involved). Following on from the technical coverage of NDE in Chapter 7 of this book, you can think of NDE knowledge as being divided into two parts.

- The NDE *techniques* themselves, that is, how to do it; this information is heavily concentrated into ASME V. In addition, API RP 577 gives useful summaries.
- NDE *acceptance criteria*: these are held in the various application codes to which equipment is constructed, for example, ASME VIII for vessels and ASME B31 series for pipework (plus many others).

16.2 Next: the NDE sections of API RP 577

API RP 577 section 9 gives useful and concise summaries of NDE. Note this important point about the API SIFE exam BoK.

- Although ASME V article 2 (RT) is not in the SIFE BoK, the RT coverage of API RP 577 *is* in the SIFE BoK.

Non-destructive examination (NDE) – including ASME V and API 577

The best approach

There is little alternative when trying to learn the content of RP 577 section 9 than to start off by reading all of it. Try to bring two objectives to your first reading.

1 See what it covers that you *don't know*. Many SIs will have good background NDE knowledge and experience so won't find the RP 577 coverage that difficult.
2 Look for typical exam questions. The API SIFE exam is all closed book, remember, so the questions, by definition, have to be fairly general.

You will need to take a general view of these sections of RP 577, because of the interlinked way in which it is set out.

Preparation for question set 16a

Question set 16a covers several NDE subjects, RT, PT, VT and MT, plus a few others. Before attempting these questions...

ARE YOU SURE?

- You know what API RP 577 (section 9) says about the *practical limitations* of RT as an NDE technique?
- You appreciate that API RP 577 (section 9) is intended to cover both in-service and new construction (SI) NDE, so you may need to filter out those statements more related to in-service inspection?

Now try question set 16a

Preparation for question set 16b

Again, question set 16b covers several different areas of NDE. If you scored less than 60% in the last question set then we recommend that you start again with your reading of RP 577 section 9, looking at the technical points it makes in more careful detail. Then, as further preparation for question set 16b, identify and mark up the following.

- Look at the figure covering hardness (Hv) testing of repair welds. Note how many Hv testing points are required, where they are located and how they are spaced out.
- How would ACFM be performed?
- What are the different types of pressure tests specified in RP 577 section 9, and why are there different types?

Make sure you look them up before attempting question set 16b.

ARE YOU SURE?

- You know what the limitations of PT testing on welds are?

Now try question set 16b

Question set 16a: API 577 section 9 (NDE)

Q1. API 577 RT density

An exposed RT film that allows 1% of the incident light to pass through has a density of:

(a) 0.1
(b) 1.0
(c) 2.0
(d) 4.0

Q2. Pressure testing API 577

Typically, pressure tests should be held for a:

(a) Minimum 30 minutes
(b) Maximum 1 hour
(c) Minimum 1 hour
(d) Minimum 2 hours

Q3. API 577

ASME V article 8 covers:

(a) ET of tubes
(b) ET of welds
(c) ET of plates
(d) ET of all components

Q4. API 577 RT density measurement

Density of RT film can be measured using a densitometer or

(a) Sensitivity meter
(b) Hole-type comparator film
(c) Step wedge comparator film
(d) Laser comparator

Q5. API 577 NDE

RT is usually not used to find and classify:

(a) Cracks
(b) Overlap
(c) Slag inclusions

(d) Porosity

Q6. API 577 PT

PT may be performed to ASME V:

(a) Article 1
(b) Article 2
(c) Article 6
(d) Article 7

Q7. API 577 VT

Visual examination of components is performed with the eyes located:

(a) In at least two different locations to get a full view of the surface
(b) At an angle of at least 30° to the surface
(c) At an angle of 30° or less to the surface
(d) Less than 6 in (150 mm) from the surface

Q8. API 577 NDE

Whose responsibility is it to establish a written material verification (PMI) programme based on API 578?

(a) The SI
(b) The M/F
(c) The inspection co-ordinator
(d) The owner/user

Q9. API 577 VT

Mechanical aids to visual examination are steel rule, thickness (feeler) gauge and:

(a) Combination set square
(b) Lighting of minimum 100 foot-candles
(c) Magnifier
(d) Borescope

Q10. API 577 MT

MT is commonly used to evaluate weld joint surfaces, intermediate checks of weld layers and:

(a) Slag inclusions

(b) Back-gouged surfaces
(c) Overlap
(d) Undercut

Q11. API 577 MT

Typical types of discontinuities that can be detected by MT are cracks, seams, laps and:

(a) Blisters
(b) Laminations
(c) Poor surface profile
(d) All of the above

Question set 16b: API 577 section 9 (NDE)

Q1. API 577 hardness testing of repair welds

How many hardness readings (impressions) in a test area on repair welds is a minimum requirement under API 577?

(a) There is no minimum requirement
(b) At least six per 1 inch (25 mm) of weld
(c) Five in 1 in^2
(d) 15

Q2. API 577 MT

WFMT is used as an MT technique for valve castings and similar because:

(a) It is more portable than other types of MT
(b) It has higher sensitivity
(c) It is more convenient to use
(d) It can be used in bright sunshine

Q3. API 577 MT

An MT pie gauge is used:

(a) On the weld with the copper side up
(b) On the weld with the copper side down
(c) Adjacent to the weld with the copper side up
(d) Adjacent to the weld with the copper side down

Q4. API 577 ACFM

ACFM is not:

(a) As sensitive as WFMT
(b) A non-contact technique
(c) Suitable for temperatures of 800°F
(d) Able to detect discontinuities through coatings

Q5. API 577 ACFM

In ACFM the probe is moved:

(a) Along the weld adjacent to the weld toe
(b) Across the weld (twice in each location)
(c) Along the weld cap itself
(d) Along the weld approximately on the interface between the HAZ and the parent metal

Q6. API 577 PT

PT shall be used instead of MT on:

(a) Plain (low) carbon steels and low-Cr alloy steels
(b) Ferritic steels
(c) Austenitic stainless steel
(d) All stainless steel

Q7. API 577 PT limitations

A limitation of using PT on welds is:

(a) It is weak at finding welding-induced defects
(b) The weld must be allowed to cool between passes
(c) It can only be used when the weld is complete
(d) All PT materials contain chemicals which can contaminate the weld

Q8. API 577 Pressure testing

The most sensitive pressure test for finding leaks is:

(a) Full hydraulic ('hydrostatic') test at $1.3 \times$ MAWP
(b) Helium test
(c) Hydrogen test
(d) Fluorescent dye test

Q9. API 577 ET

Which of these has limited use in weld inspection:

(a) ACFM
(b) ET
(c) VT
(d) WFMT

Q10. API 577 ET

A technique that gives a quick site check that components being joined by welding have the same material properties is:

(a) AUT
(b) TOFD
(c) RT
(d) ET

Q11. API 577 MT

An MT yoke placed along a circumferential weld on a vessel will identify discontinuities:

(a) In all planes
(b) Oriented circumferentially around the vessel
(c) Parallel to the weld
(d) Transverse to the weld

Chapter 17

Welding processes: API 577 and ASME IX

17.1 SIFE programme welding knowledge

The SIFE body of knowledge (BoK) incorporates a wide scope, demonstrating the importance of welding knowledge to source inspectors (SIs). Welding problems account for a large percentage of source inspection non-conformances; hence its predominance in the SIFE BoK.

Look back at the SIFE study guide (Chapter 14) and you will see that welding is covered in all of the following parts of the BoK

- API RP 577: provides good general coverage on welding processes
- AWS D1.1: specific requirements for structural steelwork
- ASME IX: welder qualification requirements
- ASME VIII and B31.3: construction (application) code sections.

This is a wide scope, split roughly 50:50 between general technical requirements (API RP 577/ASME IX) and application code requirements for specific types of equipment (ASME VIII/B31.3/AWS D1.1).

You can expect SIFE exam welding questions to be targeted more at generic welding subjects than specific parameters or data values. So this chapter will concentrate mainly on the generic welding information contained in API RP 577 and ASME IX.

First we will look at welding processes

There are four main welding processes that you need to be familiar with

- shielded metal arc welding (SMAW)
- gas tungsten arc welding (GTAW)
- gas metal arc welding (GMAW)
- submerged arc welding (SAW).

Two others that exist are flux cored arc welding (FCAW) and stud arc welding (SW), but exam questions on these are less common. As a first step, review the *Welding processes* section of API RP 577, noting what it says about each of the main welding processes. Check also what it says about them in the SIFE study guide. Next, attempt question set 17a. These are fairly straightforward questions on the welding processes.

ARE YOU SURE?

- That you can understand correctly the abbreviations SMAW, GMAW, GTAW and SAW, and that you could describe how each process works?

If so, attempt question set 17a.

Next: Welding consumables

An important part of the welding process is the use of consumables. These can be broken down into:

- filler (wires, rods, flux, coated electrodes)
- flux (and granular fluxes)
- gas (shielding, trailing or backing).

API RP 577 contains good relevant information on these. Once you are happy with which consumables relate to which welding processes then answer question set 17b, referring to API RP 577 as necessary.

ARE YOU SURE?

- That you have understood how the consumables are used in each of the welding processes?

Now attempt question set 17b.

17.2 A preliminary look at ASME IX

ASME IX is the part of the ASME BPVC that contains the rules for qualifying welding procedures and welders. It is also specifically used to qualify welders and procedures for welding to ASME VIII. Note that ASME IX is a generic code; it cannot cover all qualifications requirements, so other codes may reference it, but also contain their own qualification rules. Before progressing further, look back to

Chapter 6 of this book, which explains the role of documentation used in weld procedure and welder performance qualifications.

Formulating the qualification requirements

The actions to be taken by the manufacturer to qualify a WPS and welder are done in the following order.
Step 1: qualify the WPS.

- A preliminary WPS (this is an unsigned and unauthorised document) is prepared specifying the ranges of essential variables, supplementary variables (if required) and non-essential variables required for the welding process to be used.
- The required numbers of test coupons are welded and the ranges of essential variables used are recorded on the procedure qualification record (PQR).
- Any required non-destructive testing and/or mechanical testing is carried out and the results are recorded on the PQR.
- If all the above are satisfactory then the WPS is *qualified,* using the documented information on the PQR as the proof that the WPS works.

Step 2: qualify the welder.

The next step is to qualify the **welder** by welding a test coupon to a qualified WPS. The essential variables used, tests and results, are noted and the ranges qualified on a welder performance qualification (WPQ).

So, the steps are

- qualify the WPS and document it on the PQR
- qualify the welder and document it on the WPQ

Do not mix the rules for each of these documents or you will get confused.

17.3 The ASME IX code rules covering the WPS, PQR and WPQ

WPSs and PQRs: ASME IX QW-250

The code section splits welding variables into three groups

- essential variables
- non-essential variables
- supplementary variables.

These are listed on the WPS for each welding process. ASME IX QW-250 lists the variables that must be specified on the WPS and PQR for each process. Note how this is a very long section of the code, consisting of mainly tables, covering the different welding processes. There are subtle differences between the approaches to each process, but the guiding principles as to what is an essential, non-essential and supplementary variable remain much the same.

WPQs: ASME IX QW-350

ASME IX QW-350 lists variables by process for qualifying welder performance. These are much shorter and more straightforward than those for WPS/PQRs.

ASME IX welding documentation formats

The main welding documents specified in ASME IX have typical examples given in non-mandatory Appendix B section QW-482. The WPS and PQR formats are also shown in API RP 577 (have a look at them in Annex C). Remember that the actual *format* of the procedure sheets is not mandatory, as long as the necessary information is included. The other two that are in ASME IX non-mandatory Appendix B; the WPQ and standard weld procedure specification (SWPS) are not given in API RP 577 and so are less relevant to the SIFE exam content.

Welding documentation reviews: SIFE exam questions

The main thrust of the welding questions in the API SIFE exam is based on the requirement to understand the role of a WPS and its qualifying PQR, so these are the documents that you must become familiar with. To make it a little easier, you can assume that the review will be subject to the following limitations.

- The WPS and its supporting PQR will contain only **one** welding process.
- The welding process will be SMAW, GTAW, GMAW or SAW and will have only one filler metal.
- The base material P-group number will be either P1, P3, P4, P5 or P8.

Base materials are assigned P-numbers in ASME IX to reduce the amount of procedure qualifications required. The P-number is based on material characteristics such as weldability and mechanical properties.

Preparation for question set 17c

First step

Review API RP 577 chapter 6 *Welding procedure*, chapter 8 *Welder qualification*, and *Appendix C* for an initial overview of the subjects.

Then

Review the definitions of the WPS, PQR, WPQ, essential variables, supplementary and non-essential variables in ASME IX articles

- QW-100
- QW-200 and 251
- QW-401

Mark these sections with post-it notes.

Then

Read API RP 577 *section 6.4* in more detail.

- Section 6.4.1 lists the items that should be included on a WPS.
- Section 6.4.2 lists the items that should be included on a PQR.

Close your book and list the items you remember for a WPS, then **do the same** for a PQR.

You cannot expect ASME IX questions to be easy. Before attempting question set 17c, make sure that you have had a good look through the ASME IX articles. Don't go into too much depth; just try to get the overall picture. You will find things easier if you already have some familiarity with WPSs and PQRs.

ARE YOU SURE?

- You have followed the information on code reading and marking-up in relation to the welding sections of API RP 577 and ASME IX.

If you have, then attempt the final question set 17c

Question set 17a: Welding processes API 577

Q1. Welding processes

How is fusion obtained using the SMAW process?

(a) An arc is struck between a consumable flux coated electrode and the work
(b) An arc is struck between a non-consumable electrode and the work
(c) The work is bombarded with a stream of electrons and protons
(d) An arc is struck between a reel fed flux coated electrode and the work

Q2. API 577 welding processes

Which of the following is not an arc welding process?

(a) SMAW
(b) STAW
(c) GMAW
(d) GTAW

Q3. API 577 welding processes

How is fusion obtained using the GTAW process?

(a) An arc is struck between a consumable flux coated electrode and the work
(b) An arc is struck between a non-consumable tungsten electrode and the work
(c) The work is bombarded with a stream of electrons and protons
(d) An arc is struck between a reel fed flux coated electrode and the work

Q4. API 577 welding processes

How is the arc protected from contaminants in GTAW?

(a) By the use of a shielding gas
(b) By the decomposition of a flux
(c) The arc is covered beneath a fused or agglomerated flux blanket
(d) All of the above methods can be used

Q5. API 577 Welding processes

How is fusion obtained using the GMAW process?

(a) An arc is struck between a consumable flux coated electrode and the work
(b) An arc is struck between a non-consumable electrode and the work
(c) The work is bombarded with a stream of electrons and protons
(d) An arc is struck between a continuous consumable electrode and the work

Q6. API 577 welding processes

How is the arc shielded in the SAW process?

(a) By an inert shielding gas
(b) By an active shielding gas
(c) It is underneath a blanket of granulated flux
(d) The welding is carried out under water

Q7. API 577 welding processes

SAW stands for:

(a) shielded arc welding
(b) stud arc welding
(c) submerged arc welding
(d) standard arc welding

Q8. API 577 welding processes

Which of the following processes can weld **autogenously**?

(a) SMAW
(b) GTAW
(c) GMAW
(d) SAW

Q9. API 577 welding processes

Which of the following is a commonly accepted advantage of the GTAW process?

(a) It has a high deposition rate
(b) It has the best control of the weld pool of any of the arc processes
(c) It is less sensitive to wind and draughts than other processes
(d) It is very tolerant of contaminants on the filler or base metal

Q10. API 577 welding processes

What type of power supply characteristic is normally used with GMAW?

(a) Constant current
(b) Constant voltage
(c) Drooping characteristic
(d) Variable voltage

Question set 17b: Welding consumables API 577

Q1. API 577 consumables

In a SMAW electrode classified as E7018 what does the 70 refer to?

(a) A tensile strength of 70 ksi
(b) A yield strength of 70 ksi
(c) A toughness of 70 J at 20°C
(d) None of the above

Q2. API 577 consumables

Which of the following does not produce a layer of slag on the weld metal?

(a) SMAW
(b) GTAW
(c) SAW
(d) FCAW

Q3. API 577 consumables

Which processes use a shielding gas?

(a) SMAW and SAW
(b) GMAW and GTAW
(c) GMAW, SAW and GTAW
(d) GTAW and SMAW

Q4. API 577 consumables

What type of flux is used to weld a low-hydrogen application with SAW?

(a) Agglomerated

(b) Fused
(c) Rutile
(d) Any of the above

Q5. API 577 consumables

What shielding gases can be used in GTAW?

(a) Argon
(b) CO_2
(c) Argon/CO_2 mixtures
(d) All of the above

Q6. API 577 consumables

Which process does not use bare wire electrodes?

(a) GTAW
(b) SAW
(c) GMAW
(d) SMAW

Q7. API 577 consumables

Which type of SMAW electrode would be used for low-hydrogen applications?

(a) Rutile
(b) Cellulosic
(c) Basic
(d) Reduced hydrogen cellulosic

Q8. API 577 consumables

In an E7018 electrode, what does the 1 refer to?

(a) Type of flux coating
(b) It can be used with AC or DC
(c) The positional capability
(d) It is for use with DC only

Q9. API 577 consumables

Which of the following processes requires filler rods to be added by hand?

(a) SMAW
(b) GTAW
(c) GMAW
(d) SAW

Q10. API 577 consumables

Which of the following process(es) use filler supplied on a reel?

(a) GTAW
(b) SAW
(c) GMAW
(d) Both (b) and (c)

Question set 17c: ASME IX welding qualifications

Q1. ASME IX section QW

The conditions (including ranges, if any) under which welding must be performed are referred to in ASME Section IX as:

(a) Variables
(b) Process parameters
(c) Procedural conditions.
(d) Essential variations

Q2. ASME IX section QW

What is the main purpose for qualification of a WPS?

(a) To determine that a weldment has achieved the required properties
(b) To record the NDE requirements
(c) To determine that the welder has achieved the required standard
(d) To prove that the PQR is fit for production work

Q3. ASME IX section QW

What would be the definition of a PQR?

(a) A procedure for use in production welds
(b) A record of the welder's performance qualification
(c) A record of the welding data (variables and testing) used for a test coupon
(d) A record of the welding data for use by an approved welder on site

Q4. ASME IX section QW

What information does a PQR generally contain?

(a) The essential variables recorded during the welding of a test coupon
(b) The actual variables used and recorded during the welding of a test coupon
(c) Every variable used during the welding of a test coupon
(d) Only the essential variables used during the welding of a test coupon

Q5. ASME IX section QW

Which of the following require requalification of a WPS if they are changed outside their range?

(a) Essential variables only
(b) Essential and supplementary variables
(c) Supplementary variables only
(d) Essential, non-essential and supplementary variables

Q6. ASME IX section QG

What is an essential variable?

(a) A variable that will affect the mechanical properties of the weldment
(b) A variable that, if changed, will require requalification of a WPS
(c) A supplementary variable for a metal where notch toughness is specified
(d) A variable that can only be changed by the engineer

Q7. ASME IX section QW

What actions must be taken if a non-essential variable is changed?

(a) No actions are required, as it does not affect mechanical properties
(b) Document the change by amending or renewing the WPS
(c) Inform each WPS user personally of the change
(d) Carry out a reduced test to prove the WPS is still valid

Q8. ASME IX section QW

How would you best describe an essential variable for a performance qualification?

(a) One that will reduce the performance of a weld procedure
(b) One that affects how a welder performs his duties
(c) One that affects a welder's ability to deposit sound weld metal
(d) None of the above

Q9. ASME IX section QW

How would you best describe an essential variable for a procedure qualification?

(a) One that will affect notch toughness
(b) One that will affect mechanical properties (including notch toughness)
(c) One that becomes relevant if notch toughness is required
(d) None of the above

Q10. ASME IX section QW

Which of the following statements regarding supplementary essential variables is true?

(a) They do not affect SWPs
(b) They are the same for each process
(c) When notch toughness is required, they become essential variables
(d) Their use is determined by the code user

Chapter 18

Structural steelwork welding: AWS D1.1

18.1 Introduction

What is structural steelwork welding?

Structural steelwork comprises the large inventory of structural plates, beams, bars, bolts and similar that are used in assemblies such as plant supporting structures and frameworks, bridges, towers, chimneys, building frameworks and suchlike. Most of these are forged or rolled components made of plain carbon steel, or carbon–manganese steel, when extra strength is required, with the assemblies welded and/or bolted together using a variety of standard or special or 'high-friction' reinforced bolted joints. Welding techniques also tend to be straightforward; low-volume joints are SMAW, with high-volume production of fabricated beams and box sections normally made by SAW.

AWS D1.1 Structural welding code – steel

AWS D1.1 *Structural welding code – steel* is included in the SIFE body of knowledge. It contains many hundreds of pages, making it one of the largest welding codes available. Because of its complexity, SIFE exam questions sourced from AWS D1.1 are of a very general nature, rather than involving specific technical facts and figures about the welding of structures. Much of the general content of AWS D1.1 is similar to that covered in the ASME VIII-I sections and the welding chapters of API 577. Specific structure-related information that is in AWS D1.1 is about dimensional tolerances of components, assemblies, product marking and similar. From a practical viewpoint (rather than SIFE exam questions) some of the other codes cross-referenced in AWS D1.1 are of use to a source inspector (SI), as given below.

- American Institute of Steel Construction (AISC) code 303-10. This

FIG 18.1
API SIFE
Structural components

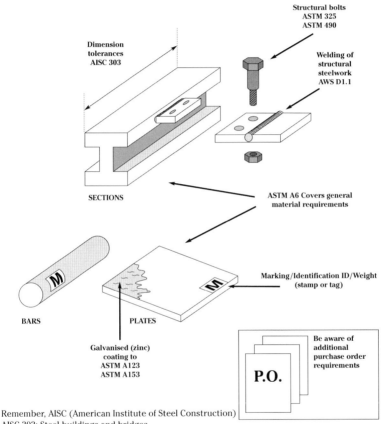

Remember, AISC (American Institute of Steel Construction)
AISC 303: Steel buildings and bridges
AISC 325: Steel construction manual
AISC 348: Structural joints using ASTM 325/490 bolts

covers engineering practice for structural steel for buildings and bridges. It contains figures of dimensional tolerances that can be useful during source inspections.
- AISC code 325: *Steel construction manual.*
- AISC code 348: *Specification for structural joints.*

Much of this information is based on simplified principles of the ASME code. Design procedures, welding procedures, assembly tolerances and similar are equally well specified as in ASME codes but do not need to be so precise in some areas, reflecting the simpler nature of the design and construction. Figure 18.1 shows a summary of the items covered.

Workmanship

AWS D1.1, like many structural codes, contains requirements that come under the general term of *workmanship*. These are distributed over several sections of the code. During a real source inspection, workmanship requirements encompass those stated on welding or assembly drawings and those of a more general nature, which can be more subjective. It is always better if these are translated into actual requirements on the inspection and test plan (ITP) to avoid as much misinterpretation as possible.

Based on the content of AWS D1.1, the following is a quick checklist of workmanship items.

- Preparation of bare metal before welding: removal of hardened, flame-cut edges and edge discontinuities.
- Weld joint fit-up: alignment, angles and uniformity.
- Dimensional tolerances: normally shown on assembly drawings, but need to incorporate allowances for weld shrinkage and distortion, so may not always be particularly straightforward.
- Surface finish prior to coating: this is generally well specified in coating standards (e.g. SSPC), but there is often quite a lot of detail that continues to be classed as 'workmanship'. Typical items are weld bead and radii blending, removal of weld spatter and preparation of hard-to-access areas. Manufacturers may take short cuts in these areas, leading to later problems with coating adhesion and breakdown.

18.2 ASTM material 'general requirements' specification

The API SIFE body of knowledge (BoK) lists three referenced ASTM material standards. They are

- ASTM SA-6: Structural rolled steel
- ASTM SA-20: Steel plates for pressure vessels
- ASTM SA-370: Material test methods.

Taken together, these three standards provide a good coverage for commonly used steels: SA-6 for structural components, SA-20 for static pressure vessels and SA-370 defining the mechanical tests that both types are subjected to. They contain about 100 pages each of specialised technical information, so do not sit well with the concept of closed-book SIFE exam questions. All three have sections on definitions – useful general information with some commonality with the contents of API 577 and the other ASME codes included in the SIFE BoK. We shall look at some useful generic points covered in these codes.

SA-6 Structural rolled steel

The full title of this document is *Specification for general requirements for rolled structural steel bars, plates, shapes and steel piling*. It covers plates, bars and other shapes such as rolled I-beams and channel sections. The first key point is that SA-6 is a 'general requirements' document that applies to 20–30 different material specifications. It is therefore not a material specification in itself but sets out general requirements for

- material thicknesses
- variability in chemical content (rather than actual values)
- surface finish of plates, bars and so on
- material marking
- dimensional tolerances on shapes (I-beams, channel sections etc.)
- mechanical testing by references ASTM A-370.

In practice, these requirements are then applied to a particular material specification under consideration, for example, A-36 *Carbon structural steel*. Note that these are not included in the SIFE BoK.

SA-20 Steel plates for pressure vessels

This is a similar 'general requirements' document to SA-6, but specifically covers pressure vessel plates rather than a long list of assorted rolled shapes. It is used across the materials range from carbon–manganese steels such as SA-516 through to higher temperature grades such as SA-515 and specific high-strength alloys. This SA-20 standard provides more details on welding than its structural steel counterpart SA-6. It specifies limits on carbon equivalent, section thickness and other parameters related to weldability. Chemical content variation is also tightly specified to keep control of hardenability.

SA-370 Material test methods

This standard can be applied to all steel products and is cross-referenced in just about every ASTM and ASME code of relevance to the source inspector. It contains detailed step-by-step information on performing all the common mechanical tests, plus definitions and background information on the mechanical properties themselves. They include

- tensile (strength) tests
- Charpy/Izod impact (toughness) tests
- hardness (surface indentation) tests.

As a source inspector, generalised coverage of these topics is given in API 577 *Welding and metallurgy*, and in construction codes themselves, so SA-370 really only needs to be consulted for specialised technical detail on test methods – when witnessing mechanical tests in a test lab. for example. At the manufacturing stage, SA-370 gives essential information on *where* test pieces are cut from in various manufactured components. This is important, as specimen origin and orientation can have a significant effect on the grain size and direction, thereby affecting the test results. Charpy impact tests on thick components can be the most variable with a difference of 30–40% possible between test piece locations.

Try this next question set 18a. These are awkward questions sourced from AWS D1.1, so you will need to track down the answers 'open book' using the code.

Question set 18a: AWS D1.1: Structural welding

Q1. Welder qualification AWS D1.1

Can a welder, qualified under ASME IX, weld structures manufactured in accordance with AWS D1.1?

(a) If the positions and ranges are suitable
(b) If he was qualified within the previous 6 months
(c) No
(d) With the inspector's approval

Q2. Scope AWS D1.1

AWS D1.1 is applicable to:

(a) All welded fabrications
(b) Pressure vessels
(c) Structures
(d) Pressure piping

Q3. Visual inspection AWS D1.1

You are inspecting the finished weld on a tubular TKY connection, 25 mm wall thickness, length 850 mm, specified weld size 35 mm. in ASTM A517 material. The weld has cooled to ambient temperature having been completed over the preceding night shift. You find the weld to be undersize by a maximum of 2.5 mm over a cumulative length of 90 mm. The weld is:

(a) Acceptable
(b) Unacceptable, the weld undersize exceeds code requirements
(c) Unacceptable, the length of the undersized weld exceeds the code
(d) Unacceptable, undersized welds are not permitted on tubular connections

Q4. Visual inspection AWS D1.1

You are inspecting the finished weld on a tubular TKY connection, 25 mm wall thickness, length 850 mm, specified weld size 35 mm in ASTM A517. The weld has cooled to ambient temperature having been completed over the preceding night shift. You find the weld to be undersize by a maximum of 2.5 mm over a cumulative length of 90 mm. As the inspector, you should:

(a) Postpone the inspection until the following day
(b) Reject the weld
(c) Accept the weld
(d) None of the above

Q5. WPS qualification AWS 1.1

A WPS test weld for a TKY joint is being made on an 8 in NB pipe. What is the allowable difference in the I.D. of the test members?

(a) 5 mm minimum
(b) 10 mm minimum
(c) 10 mm maximum
(d) 5 mm maximum

Chapter 19

Pipework ASME B31.3

The SIFE body of knowledge (BoK) specifically includes selected parts of the ASME pipework code ASME B31.3 *Process piping*. This is the most commonly used of the ASME B31 series codes, being in worldwide use in oil and petrochemical industries both inside and outside the USA. ASME B31.3 is a document containing many construction options, based mainly on differing fluid classes, anticipated levels of risk, and therefore integrity required.

19.1 The SIFE BoK content

The SIFE BoK is limited to five chapters of ASME B31.3 as shown below. They cover quite a wide scope of manufacturing inspection and testing activity, but specifically exclude design (Chapter II) calculations (these are not in the SI's scope of responsibility).

   B31.3 Ch 1: *Scope and definition*
   B31.3 Ch III: *Materials*
   B31.3 Ch IV: *Standards for piping components*
   B31.3 Ch V: *Fabrication*
   B31.3 Ch VI: *Inspection, examination and testing*

Types of questions

As discussed in several chapters of this book, API closed-book exam questions are based around general principles of code content rather than specific clause-by-clause detail. This principle is particularly relevant to ASME construction codes, which contain almost limitless amounts of technical detail. Practically, many exam-style questions raised from ASME B31.3 will be similar to those from ASME VIII-I, the principles being similar. This is a useful point to bear in mind when reading through the relevant code sections. As always, be on the lookout for question points based on

FIG 19.1
The scope of ASME B31.3

B31.3 COVERS PIPEWORK OUTSIDE PROCESS UNITS

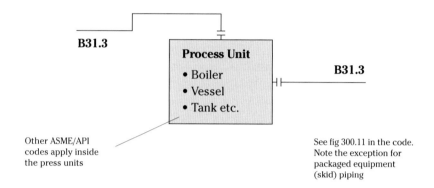

FLUID/PIPING CATEGORIES ARE (SEE 300.1)

- High-purity fluid service
- High-pressure fluid service
- Elevated temperature fluid service (see Table 302.35 note b)
- Cat M fluid service: toxic
- Normal fluid service: all fluids not covered by other categories
- Cat D: < 150 psi, -20°F to 266°F non-damaging.

See the definitions section (300.2) ASME B31.3, Chapter 1

FIG 19.2
Heat treatment definitions
B31.3 (300.2) Chapter 1

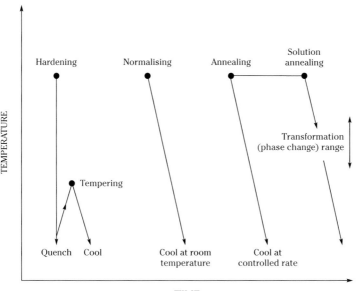

Mechanical properties are heavily influenced by
- The temperature reached (and exposure time)
- The cooling rate

These can produce very different mechanical properties in materials of identical chemical composition.

- **responsibilities and authorities**: who is responsible for what
- **code scope**: which components a code covers, and where its jurisdiction ends (see Figure 19.1)
- **general engineering principles** of inspection: which parts an SI would actually look at
- **visual, NDE and pressure testing**: the scope and extent of testing are particularly important here.

Engineering definitions feature heavily in chapter 1 of ASME B31.3 (they are in section 300.2). There are more than 100 listed, ranging from general metallurgical definitions about welding and heat treatment (similar to those you can find in API 577) to more piping-specific ones about fluid/construction categories and pipework component types. Figure 19.2 shows some important definitions for heat treatment processes.

19.2 ASME B31.3 quick reference points

As the API SIFE examination comprises all closed-book questions, you will not be required to memorise lots of detailed points from the chapters of ASME B31.3. You need to read the relevant chapters, but the best way to do this is in conjunction with Figures 19.1–19.3. These contain important learning points that are suitable for closed-book exam questions. In reviewing these figures, look for the origin of the points raised in the body of the code pages. After reviewing each, try the short question set that follows, looking up the source of the answer in the code if you get any wrong.

ASME B31.3 Chapter III Materials

Chapter III follows a pattern common to other ASME construction codes in controlling which materials can and cannot be used to construct B31.3 pipework. As a short chapter containing a lot of specific graph and table data unsuitable for SIFE closed-book exam questions, the main exam-related content is restricted to points of general principle. Overall, most of these would be common to other fixed equipment items such as vessels, valves or tanks, or covered in the more general materials/metallurgy content of API RP 577 (see Chapter 5 of this book). Figure 19.3 shows some principles that are important.

Extensive details of the allowable stress values for pipework are listed in Appendix A (Table A-1) of B31.3. This is essential design information

FIG 19.3
ASME B31.3 Chapter III: Materials

MATERIALS ARE ACCEPTED IF THEY ARE:

ALL IMPORTANT MATERIAL IS:

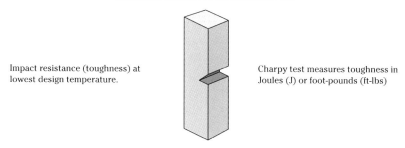

Impact resistance (toughness) at lowest design temperature.

Charpy test measures toughness in Joules (J) or foot-pounds (ft-lbs)

Graphs and tables specify when Charpy tests are required.

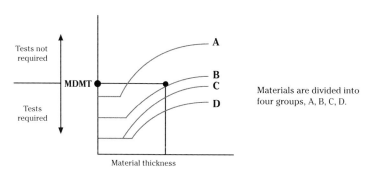

Materials are divided into four groups, A, B, C, D.

Many low-carbon steels are considered suitable for use down to a temperature of -20°F (-29°C) but not relevant to the SIFE BoK. Note a couple of general principles below that are relevant, however.

- **Temperature limitations**: construction materials have a maximum and minimum temperature at which they can be used. At a temperature that is too high, they lose strength. At low temperature they become brittle and liable to crack.
- **Material certificates** contain laboratory test results that may need to be reviewed by the SI.
- There are many supporting standards covering valves and other piping components (see the long list in B31.3 table 326.1).

ASME B31.3 Chapter V Fabrication

Fabrication details are difficult to use as a source of closed-book, text-only exam questions, owing to their reliance on diagrams. In addition, fabrication of pipework is limited mainly to longitudinal, circumferential and branch butt welds, thereby limiting the level of technical detail that is necessary. Practically, there is little exam-relevant material in this chapter that is not covered in much more detail in API RP 577. There is also commonality with the fabrication chapter of ASME VIII. Have a look at the following points, however, for SIFE exam revision purposes.

Q: Are WPS/PQR/WPQs required for pipework?
Ans: Yes, as with all coded pressure equipment items (see B31.3 section 328).

Q: Are there any specific terms and definitions of interest?
Ans: Yes, make sure you know the meanings of

- weld misalignment
- effective throat (weld) dimensions
- weld *size*
- backing strip
- slip and socket flange welds.

Q: What are the different reasons for weld preheat and post-weld heat treatment (PWHT)?
Ans: Preheat reduces thermal gradients whereas PWHT reduces residual stresses and refines the metal's grain structure (see B31.3 sections 330 and 331).

Q: What factors determine if preheat and PWHT are required for a specific pipework fabrication?
Ans: Material P-number (a weldability grouping), tensile strength and

nominal wall thickness (have a look at the two heat treatment tables 330.1.1 and 331.1.1).
Q: Is it possible to tell when PWHT has not been done properly?
Ans: Yes, to a limited extent, using hardness tests (see 331.1.7).

ASME B31.3 Chapter VI: Inspection, examination and testing

This ten-page chapter contains a lot of important information relevant to source inspections. In line with US practice, the inspection witnessing activities are directed at the ASME National Board commissioned authorised inspector (AI) role, but for exam purposes you can assume these are the same as the API SI, if they are performing a similar function.

Starting with NDE, the chapter specifies the extent (i.e. percentage of welds) to be examined and the defect acceptance criteria (DAC) to be used to check indications for B31.3 code compliance. DAC levels are found in two separate locations: table 341.3.2 for all indications excluding linear indications found by UT and clause 344.6 for linear indications themselves. Note some key points about DAC.

- They differ between types of weld (longitudinal, girth and so on).
- They differ between pipe/fluid class (normal, Cat D, severe cyclic and so on).
- Cracks are unacceptable in all situations.
- UT DAC are dependent on material thickness.

Pressure testing
ASME B31.3 uses its own terminology for types of pressure test, differing in some areas from those used in ASME VIII and related API codes. Interpret the B31.3 definitions as described below.

- **Leak testing** (clause 345.4.1) means *full hydraulic testing*, using liquid. Minimum test pressure = 1.3 × MAWP × temperature correction factor.
- **Pneumatic leak testing** (clause 345.5) means *full pneumatic testing* using air or inert gas. Minimum test pressure = 1.1 × MAWP (no temperature correction required).
- **Sensitive leak testing** (clause 345.8) is a *low-pressure bubble test*, testing for leaks rather than strength. Minimum test pressure is 25% MAWP or 15 psi, whichever is the lower.

Questions on pressure testing are common in the API SIFE exam. Expect them to be fairly general in nature with straightforward answer options on test types, test pressure or safety aspects. Pneumatic tests are always the most dangerous owing to the large amounts of stored energy held in compressed air or gas. Note that pressure testing of valves is covered by a separate document, API 598 *Valve inspection and testing* (see Chapter 21 of this book). This contains much more technical detail on the actual procedures of pressure testing. It also covers measurement of leak rates.

To progress your learning on ASME B31.3, attempt the following question set 19a. Before attempting it, read carefully the scope and definition section of the code. Make sure also that you understand

- the different types of fluid service
- which parties are responsible for what, within the boundaries of the code content.

ARE YOU SURE?

- You know what the different types of fluid service are?
- You know the temperature and pressure limits for Cat D fluid service?

If so, attempt question set 19a.

Question set 19a: B31.3, Chapter 1

Q1. B31.3 Definition

Seamless pipe can be produced by a process of piercing in a billet then:

(a) Casting
(b) Welding
(c) Rolling
(d) Resistance welding, with no filler metal

Q2. B31.3 Definition

The exposed surface of a weld on the side from which the welding is done is called:

(a) The cap
(b) The land
(c) The reinforcement
(d) The face

Q3. B31.3 Weld size definition

A fillet weld with unequal leg sizes of $\frac{1}{2}$ in and $\frac{5}{8}$ in is described as a weld of 'size':

(a) $\frac{3}{8}$ in
(b) $\frac{1}{2}$ in
(c) $\frac{9}{16}$ in
(d) $\frac{5}{8}$ in

Q4. B31.3 Definition

A pipework system carrying fluid to which a single exposure can cause serious harm to people should be constructed to ASME B31.3:

(a) Category D rules
(b) Category M rules
(c) Category X rules
(d) Category LS (lethal service) rules

Q5. B31.3 Definition

A person who welds using a semi-automatic welding machines is:

(a) A welder

(b) A welding operator
(c) Not required to be qualified
(d) A semi-automatic welder

Q6. B31.3 Scope and definition

Who is responsible for providing components and workmanship in compliance with the requirement of a construction code (e.g. B31.3)?

(a) The manufacturer
(b) The designer
(c) The owner's inspector
(d) All the above parties

Q7. B31.3 Definition

A weld joining the surfaces at right angles to each other in a lap joint or tee joint is a:

(a) Cruciform joint
(b) Partial penetration weld
(c) Fillet weld
(d) Tee weld

Q8. B31.3 Fluid service

Toxic fluid would most commonly be classed by B31.3 as:

(a) Category D
(b) Category M
(c) Category T
(d) Category X

Q9. B31.3 Fluid service

A 'normal' fluid under B31.3 is:

(a) Any fluid that is non-corrosive
(b) Any fluid that is at less than 150 psi
(c) Any fluid that has low 'stored energy'
(d) Any fluid that does not fit into any of the other fluid categories of B31.3

Q10. B31.3 Fluid categories

Who is responsible for deciding what B31.3 fluid category (e.g. D, M, high pressure etc.) a piping system is designed to:

(a) The designer
(b) The owner/user
(c) The fabricator
(d) It is specified in B31.3, Chapter 1

Chapter 20

Pressure vessels: ASME VIII-I

20.1 ASME VIII-I clauses

The SIFE body of knowledge (BoK) specifically includes selected parts of the ASME VIII Division 1 *Rules for construction of pressure vessels*. ASME VIII-I is a long and complex code with source inspection-related content spread across several different subsections: UW (welding), UG (general) and UCS (carbon steel). The following is a list of those subsections included in the SIFE BoK.

- Definitions in Appendix 3
- Materials: UG-4 to UG-15
- Fabrication: UG-75 to UG-85
- Inspection and testing: UG-90 to UG-103
- Marking and reports: UG-115 to UG-120
- Welding general: UW-1 to UW-3
- Welding materials: UW-5
- Fabrication: UW-26 to UW-42
- Inspection and tests: UW-46 to UW-54
- Marking and reporting: UW-60
- PWHT: UCS-56
- RT examination: UCS-57

This is a larger BoK list than that for ASME B31.3 pipework, reflecting ASME VIII's greater emphasis on welding, marking and reports. Vessels also contain more welding and fabrication issues than pipework, resulting in more code clauses relevant to the SIFE BoK.

ASME VIII-I materials UG-4 to UG-15

From a practical viewpoint much of the content of the ASME VIII-I code clauses is similar to that included in the SIFE study guide (see

FIG 20.1
ASME VIII-1 materials
UG-4 to UG-15

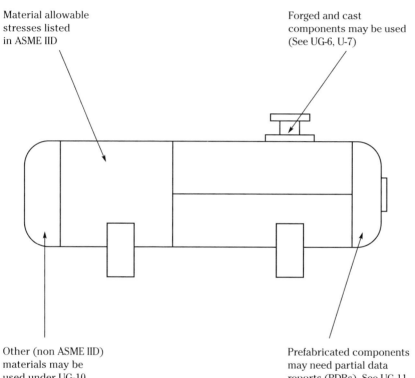

Material allowable stresses listed in ASME IID

Forged and cast components may be used (See UG-6, U-7)

Other (non ASME IID) materials may be used under UG-10 recertification procedure

Prefabricated components may need partial data reports (PDRs). See UG-11

FIG 20.2
ASME VIII-I fabrication requirements: UG-75 to UG-85

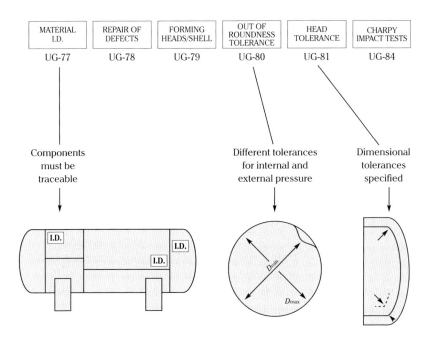

ASME VIII-I has carefully specified multiple tolerances of head shape. The objective is to avoid *excessive bending stress* in these components.

Pressure vessels: ASME VIII-I

Chapter 14 of this book) and other ASME codes. Figure 20.1 shows easily accessible points that contain an angle suitable for exam questions. It is advisable not to go into much more depth than this when looking at UG-4 to UG-15. Much of this is component-specific information which is just too complicated for SIFE exam questions.

ASME VIII-I fabrication UG-75 to UG-85

Do not confuse fabrication requirements with *design*. Pressure equipment design methods and calculations are not included in the SIFE BoK, whereas fabrication details *are*, as they sit squarely within the remit of the source inspector (SI). Code clauses UG-75 to UG-85 provide concise coverage of a manageable number of key fabrication principles, as shown in Figure 20.2. Note how these lead chronologically through the fabrication process. Do not get too involved with UG-84 *Impact tests* – it contains specific information which is not suitable for closed-book exam questions. Concentrate on the nature of impact (Charpy) tests and why they are required, but do not get caught up in the detail of the graphs and tables.

ASME VIII-I inspection and testing UG-90 to UG-103

These clauses relate directly to the *role* of the SI (or National Board authorised inspector (AI)). The first one UG-90 is a 'summary clause', which does little more than replicate a list of other code clauses that have be checked during the vessel inspection. It is a good checklist for the SI, but not a particularly easy source of SIFE closed-book exam questions. From an examination viewpoint most of the detailed points on inspection and testing hidden away in UG-90 to UG-97 are covered in Part A of this book. Perhaps the most important ASME VIII-specific points are these covering pressure testing.

ASME VIII-I marking and reports: UG-115 to UG-120

These parts of ASME VIII-I are some of the most ASME-specific parts of the SIFE BoK. They are almost exclusively related to ASME VIII vessels and have little relevance to other components or practices used in other countries' codes and standards. Strictly, they are included in ASME VIII-I for the guidance of the National Board AI, rather than the more generalist SI, who would not hold a National Board commission. SIFE exam questions are sourced from these clauses, but they are necessarily of a simplified nature, to reflect the role of the SI in

FIG 20.3
ASME VIII-I marking and reports UG-115 to UG-120

NAMEPLATE AND DOCUMENTATION REQUIREMENTS

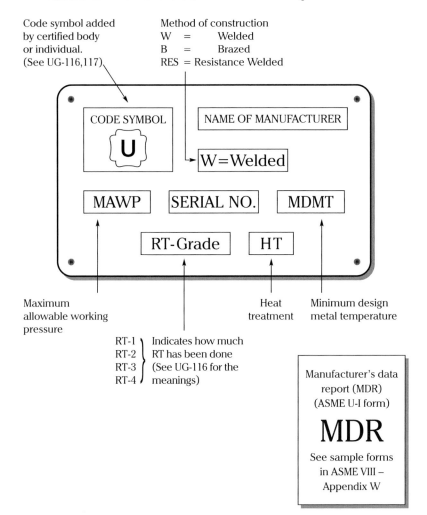

being involved in inspection of various types of static plant, not just ASME VIII-I vessels.

Figure 20.3 summaries the main points you need to know from UG-115 to UG-120. It is good practice to track these down in the actual code clauses to see where they come from – but do not get involved in too much detail. The vessel nameplate supplies outline information, whereas the manufacturer's data report (MDR) contains fuller engineering detail on material specification, thicknesses, heat treatments, NDE and test pressures. The MDR and nameplate are produced by the vessel manufacturer but may only have the code 'U' stamp applied by a certified inspection body or individual.

How important are vessel markings?
Vessel nameplate/markings and their accompanying MDR are given great importance by the ASME code. In isolation, they do not contribute to the physical integrity of the vessel but do give a degree of comfort that all the required code-compliance activities listed in UG-90 have been followed. As an API SI, it is your role to check the overall completeness of the nameplate/markings and validating documentation rather than understand each step in great depth. That is the job of the manufacturer and the AI.

ASME VIII-I fabrication details UW-26 to UW-42

Despite the apparent detail of UW-26 to UW-42, the fabrication section of ASME VIII-I contains straightforward information, most of which could apply equally to any welded pressure equipment component. Overall control of welding is achieved by a mandatory requirement for WPS/PQR/WPQ certification to ASME IX. This is supplemented by common-sense guidance on the weld-prepared plate edges and similar for the completed welds. Figure 20.4 shows a summary. Expect SIFE exam questions on this to be straightforward. The guidance given in Chapter 6 of Part A of this book is also relevant.

Finally, the RT requirements of ASME VIII-I: UW-51 to UW-52

RT is an NDE technique favoured heavily by the ASME VIII-I code. UT can be used as a valid alternative, if you read the small print, but most of the technical requirements of the code are still written around the use of RT.

FIG 20.4
ASME VIII-I vessel fabrication details
– UW-24 to UW-42 –

General principles: WPS/PQR/WPQs are mandatory (UW-28)

WPS	PQR	WPQ
Ensures welds meet code requirements	Proves that the WPS 'works'	Proves the welder can weld to the WPS (Follow ASME IX)

PLATE CUTTING, FITTING ALIGNMENT: UW-31

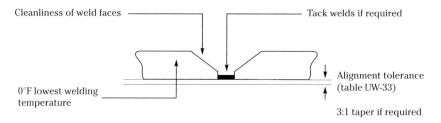

- Cleanliness of weld faces
- Tack welds if required
- 0°F lowest welding temperature
- Alignment tolerance (table UW-33)
- 3:1 taper if required

FINISHED WELDS: UW-35

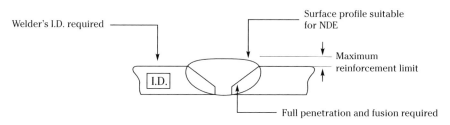

- Welder's I.D. required
- Surface profile suitable for NDE
- Maximum reinforcement limit
- Full penetration and fusion required

HT | **UW-40**   PWHT to UW-40 (and tables in UCS-56)

Mandatory RT

Table UCS-57 is in the SIFE BoK and gives a clear list of material thicknesses above which RT of butt welded joints is mandatory. This depends on material P-number, as shown simplified and tabulated below.

P-number (group)	Thickness above which RT is mandatory
P1 (all)	$1\frac{1}{4}$ in (32 mm)
P3 (all)	$\frac{3}{4}$ in (19 mm)
P4 (1, 2)	$\frac{5}{8}$ in (16 mm)
P5 (1, 2)	0 (i.e. for all thicknesses)
P9 (1)	$\frac{5}{8}$ in (16 mm)

The principle behind this table is simple, the higher the P-number, the greater the hardenability of the material, and so the greater the risk of cracking as section thickness increases. RT is done to find any cracks that have occurred, with acceptance criteria set out in the body of clauses UW-51 and UW-52. Salient points are listed below.

- Cracks, lack of fusion (LOF) or lack of penetration (LOP), are not allowed for either spot or full RT welds.
- Linear (including aligned) indications such as slag are acceptable within limits.
- Rounded indications are assessed for full RT welds (see VIII-I Appendix 4) but are not an issue for spot RT welds.

Checking the extent of RT and assessing the results against code are common duties of an SI. Owing to the code-specific nature of the defect acceptance criteria, this does not translate easily into SIFE closed-book exam questions. Instead, questions are more likely to concentrate on responsibilities and authorities related to the RT activity, supplemented by generic question that could also apply to other welded components, as well as ASME VIII-I vessels.

To start your learning of ASME VIII, attempt the 15 questions in the following question set 20a. This covers a wide-ranging selection of code clauses, so you will need to track down the answers in ASME VIII if you want to get them right.

ARE YOU SURE?

You have reviewed the ASME VIII-I code clauses covering

- material identification and recertification
- impact testing
- head and shell tolerances.

If so, then attempt question set 20a.

Question set 20a: ASME VIII-UG

Q1. UG Bolts and studs

When can bolts and studs be used for the attachment of removable pressure retaining parts to ASME VIII-I pressure vessels?

(a) Always, as long as ASME VIII-I code clauses are complied with
(b) Only for non-lethal service vessels
(c) Only if the owner/user and inspector give approval
(d) Never. ASME VIII-I requires pressure-retaining components to be welded

Q2. UG Material identification

Material traceability during manufacture of an ASME VIII-I vessel should be maintained by:

(a) The manufacturer
(b) The manufacturer and SI
(c) The manufacturer, SI and ASME authorised inspector (AI)
(d) The AI only

Q3. UG Impact tests. Test specimen size

What are the dimensions of a standard-size impact (Charpy) specimen?

(a) $50\,\text{mm} \times 8\,\text{mm} \times 8\,\text{m}$
(b) $2.165\,\text{in} \times 0.394\,\text{in} \times 0.394\,\text{in}$
(c) $155\,\text{mm} \times 8\,\text{mm} \times 8\,\text{mm}$
(d) $2\,\text{in} \times 10\,\text{mm} \times 10\,\text{mm}$

Q4. VIII-I Defects in raw material

If manufacturing (mill) defects are found in new material destined to be incorporated into an VIII-I vessel before presenting it to the AI:

(a) They may be repaired by the manufacturer before presenting it to the AI
(b) They may be repaired if approved by the AI
(c) The AI should specify additional tests and/or NDE
(d) It should be rejected by the AI and returned to the manufacturer

Q5. VIII-I Head tolerances

On an ASME VIII-I vessel, when may a transition taper be machined on the head (of thickness T) or shell (of thickness t) to make it a 'good fit'?

(a) It may not; the material must be the same thickness
(b) When it has a taper of a least $6T$
(c) When it has a taper of at least $3t$
(d) When it has a taper of at least $3(T - t)$

Q6. VIII-I Material suitability

The suitability of materials used for ASME VIII-I vessels should be decided by:

(a) The user
(b) The SI
(c) The vessel engineer
(d) All of the above

Q7. VIII-I Welding materials

ASME VIII-I pressure vessels shall conform with:

(a) ASME IIC
(b) ASME II
(c) ASME V
(d) ASME XI

Q8. VIII-I Recertification of other cert material

When may material that is fully identified but does not comply with an ASME II specification be used in an ASME VIII vessel construction?

(a) When it has been recertified by the manufacturer
(b) When the owner/user agrees
(c) When the owner/user and inspector agree
(d) It may not

Q9. VIII-I Cutting plates

Plates for ASME VIII-I vessels cut by oxygen or arc cutting:

(a) May not be used for fabrication
(b) May be used after mechanically grinding the cut edges
(c) May not be used after mechanically grinding the cut edges

(d) May be used after additional heat treatment

Q10. VIII-I Materials

Materials suitable for pressure components in ASME VIII-I vessels may be listed in ASME II and:

(a) Meet only one material specification
(b) Meet more than one material specification
(c) Not be used for non-pressure parts
(d) Are restricted as to the method of production

Q11. VIII-I Material marking

During manufacturer of an ASME VIII-I vessel, marking of pressure part material:

(a) May be marked by any method agreed by the AI
(b) Must be hard-stamped using low-stress stamps
(c) Must be hard-stamped or chemically etched
(d) Must be carried out by the AI

Q12. VIII-I Material marking

When an ASME VIII-I vessel head is contracted out by the manufacturer to a sub-manufacturer and the original material markings are unavoidably cut out, who should transfer the markings to a new location on the head?

(a) The AI
(b) The original vessel manufacturer
(c) The sub-manufacturer making the head
(d) The AI in conjunction with the original vessel manufacturer

Q13. VIII-I forgings and castings suitability

Cast material:

(a) May be used for construction of VIII-I pressure vessels
(b) May only be used in VIII-I pressure vessels if the material has been worked sufficiently to remove the coarse ingot structure
(c) May not be used for pressure retaining components in VIII-I pressure vessels
(d) Shall not be used at all in VIII-I pressure vessels; it is too brittle

Q14. VIII-I Forming of shells and heads (OOR)

The maximum out of roundness of an VIII-I vessel shell under construction should not exceed:

(a) 1% of the nominal diameter (must be measured on vessel ID)
(b) 1% of the nominal diameter (must be measured on vessel OD)
(c) 1% of the nominal diameter (must be measured on vessel ID or OD)
(d) 2% of the inside diameter

Q15. VIII-I Formed head tolerances

For ASME VIII-I elliptical formed heads, for a vessel with nominal inside diameter D, a maximum allowable dimensional deviation is:

(a) $+/- 5/8$% Di on the inside of the head
(b) $+/- 1/4$% Di on the inside of the head
(c) $+/- 1$% Di on the inside of the head
(d) $+/- 1$% Di on the outside of the head

Chapter 21

Valves and testing

As a source inspector (SI) you will inspect a lot of valves. There are thousands of them in any process plant, with a wide variety of types, sizes and specifications. Two main codes govern the world of valve manufacture and inspection, as follows.

- **ASME B16.34** *Flanged, threaded and welding end valves*. This covers the design and manufacture of valves in common cast materials. It categorises them into standard and special class, linking each to acceptable pressure–temperature ranges. It also specifies the extent of NDE and the usual requirements for marking and documentation.
- **API 598** *Valve inspection and testing*. This follows on from ASME B16.34, covering the testing of common types of valves both during shop manufacture and in service. It describes the different types of tests with detailed information on procedures, test pressures and acceptable leakage rates.

21.1 SIFE BoK content

Despite its importance in the valve industry, the construction code ASME B16.34 is *not* in the SIFE BoK. Instead the BoK concentrates on API standard API 598. This is of greater relevance to the SI and the various pressure and leakage tests that are commonly witnessed in a manufacturer's works.

Owing to the detailed numerical information that is included in API 598, there is a limit to the content that can realistically be translated into SIFE closed-book exam questions. Viable subjects for questions are

- **terms and definitions** given in chapter 3 of API 598
- the **objectives and types** of pressure tests performed on valves; this information is distributed over several of the other chapters of API 598
- general information on **valve types** (gate, plug, ball, globe, check and

so on). This may be considered general engineering knowledge that SIs should have, despite the fact that the main document illustrating the types (API RP 574 *Inspection practice for piping system components*) is not included in the SIFE BoK. There are no pictures of these types in API 598, as there are in API RP 574.

Other numerical information such as test pressures, durations and leakage rates is well supplied in API 598 but is unsuitable for use as closed-book exam questions.

21.2 Types of valve tests

There are four main types of pressure tests in common use during the manufacture of metallic valves

- shell (or body) test
- backseat test
- low-pressure closure test
- high-pressure closure test.

Any of these can be specified to be performed on new valves, or following the repair of valves that have already been in service. For tests on individual valves, the test configurations are usually done normally using bolted flange connections. For mass production items, most manufacturers will have an automated test rig using hydraulic closures, enabling tests to be performed in quick succession. The tests are done as follows (see Figure 21.1).

The shell test

This is a hydro test of the valve body using water or a light, low-flammability oil such as kerosene. Blanks enclose the inlet and outlet flanges and the valve bonnet is either removed and blanked off, or left in place with the valve positioned half-open so pressure can access both sides. The test provides a check of the resistance of the valve body to brittle fracture and major leakage due to casting defects and so on. Note that it may not be a 100% test of whether the valve body will leak in service, particularly if it is destined for gas, a vapour or steam service.

The backseat test

This is only used for gate and globe valve types that have a *backseat design*. This is a design feature in which, when the valve is screwed fully

FIG 21.1
Valve testing: API 598

SHELL (BODY) TEST

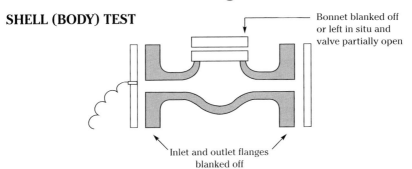

Bonnet blanked off or left in situ and valve partially open

Inlet and outlet flanges blanked off

BACKSEAT TEST

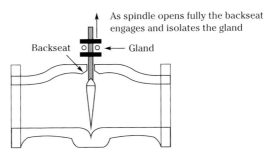

As spindle opens fully the backseat engages and isolates the gland

Backseat — Gland

CLOSURE (LEAK) TEST

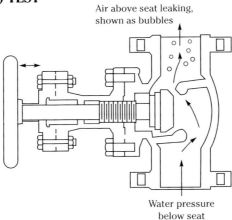

Air above seat leaking, shown as bubbles

Water pressure below seat

See API 598 for full test details.

open, a small wedge-shaped seal between the gland and the inside of the body is closed, effectively isolating the gland from the fluid pressure. This is useful for adding or lubricating gland packing without depressurising the system. For safety reasons, not all valves or fluids are suitable for this.

The backseat test is done (see Figure 21.1) by pressurising the valve body, then simply screwing up the spindle until the backseat engages. The test pressure used can vary depending on customer requirements, but it is sometimes less than that used for the shell test. API 598 contains a table showing the pressures required for different types of valves.

Low- or high-pressure leak (closure) test

These are separate tests used to check whether the valve leaks across the seat and they differ only in the pressure used. For a high-pressure leak test, the medium used is water, for safety reasons, and leakage across the seat is measured in drops. Low-pressure leak tests use air or gas on one side with water on the other, so leakage is measured by the number of bubbles observed. This is a more searching test than with water, gas being much better at finding small leaks.

Now try question set 21a covering valve types and tests.

API SIFE question set 21a: Valve inspection and testing: API 598

Q1. Valve codes

A visual quality standard for cast valve bodies and other components is:

(a) MSS-SP-45
(b) MSS-SP-55
(c) MSS-SP-91
(d) FCI 70-2

Q2. Valve inspection scope

Surface and/or volumetric NDE bodies should be performed:

(a) Only if specified in the purchase order
(b) On a 10% sample of valve cast components
(c) On all valve bodies only
(d) On all valve bodies, bonnet and cover castings

Q3. Valve testing types API 598 backseat test

A pressure test performed by applying pressure inside an assembled valve with the valve fully open is a:

(a) Low-pressure closure test
(b) Backseat test
(c) Shell test
(d) High-pressure pneumatic shell test

Q4. Valve testing: API 598

A pressure test performed by applying pressure inside a partially open assembled valve with the ends closed and gland fully tightened is a:

(a) Low-pressure leak test
(b) Backseat test
(d) Shell test
(c) Closure test

Q5. API 598 Valve testing: Backseat test

When can the backseat test and shell test be combined on a gate or globe valve?

(a) For valves <NPS 4 when volumetric devices are used
(b) For valves ≤NPS 4 when volumetric leakage detection devices are used
(c) For valves ≤NPS 4 and test is performed at lower than 100°F (38°C)
(d) It is not possible; they are different tests

Q6. General: Valve types

A valve in which the flow path inside the valve body changes twice by 90 degrees is a:

(a) Gate valve
(b) Butterfly valve
(c) Plug valve
(d) Glove valve

Q7. Common valve code

Flanged, threaded and welding-end valves are commonly made to:

(a) ASME B16.11
(b) ASME B16.5
(c) ASME B16.34
(d) API 598

Q8. Valve pressure testing API 598

Shell testing is required for:

(a) All valves
(b) All valves excluding check (non-return) valves
(c) All valves with a 'backseat' feature
(d) Ensuring the valve will not leak by ('pass') in service

Q9. Valve pressure testing API 598

A valve shell test is performed at:

(a) Maximum working pressure
(b) Above design pressure
(c) Design pressure
(d) As low on pressure as flexible for safety purposes

Q10. API 598 Valve pressure testing

A valve shell test is usually:

(a) Hydro test
(b) Pneumatic test
(c) Low-pressure leak (closure) test
(d) High-pressure leak (closure) test

Q11. API 598 Valve pressure testing

API 598 provides details on shell tests on cast valves if:

(a) Minimum shell test pressure
(b) Maximum shell test pressure
(c) Maximum shell test duration time
(d) Maximum allowable shell test leakage rates

Q12. API 598 Valve pressure testing fluid

Valve shell tests shall not be performed using:

(a) Air
(b) Water above 40°C
(c) A liquid with a viscosity higher than water
(d) Kerosene (gas oil)

Q13. API 598 LP closure testing

When witnessing a low-pressure closure test on a check valve, the pressure shall be applied:

(a) Between the valve seats
(b) On the downstream side
(c) On the upstream side
(d) On both downstream and upstream side in turn

Q14. API 598 LP closure testing

When witnessing a low-pressure closure test on a check valve, the pressure shall be applied:

(a) Successively to each side of the closed valve with the other side open to atmosphere
(b) Only from the downstream side
(c) Only with the pressure under the seat

(d) Between the seats and the bonnet area, to check for leaks from both sides

Q15. API 598 Pressure testing of gate valves

When closure testing a gate valve the test pressure should be applied:

(a) On the upstream side only
(b) Between the seats and the upstream side simultaneously
(c) Successively to each side of the closed valve with the other side open to atmosphere
(d) Between the seats only

Chapter 22

Material verification (API RP 578)

Material verification is alternatively known as positive material identification (PMI). It refers to the activity of verifying that the materials used in a component are actually the correct ones that have been specified. Similarly, it is used to verify that the chemical analysis information shown in a material test report (MTR) is a true and correct representation of what the material actually *is*. Material verification has become increasingly important in recent years as increasing amounts of low-quality 'rogue materials' have found their way onto the market. This makes it an essential activity of the source inspector (SI).

22.1 The SIFE BoK content

The ASME construction codes VIII-I/B31.3 address the need for verifiable material traceability for pressure equipment components but predominantly do this by relying on the accuracy of the information shown in MTR and the manufacturer's data report (MDR). To complement this with specific information on MTR verification, the SIFE BoK includes the content of API RP 578 *Material verification programmes for new and existing alloy piping systems*. This also covers (as the title suggests) verification of materials of systems that are already completed and in service, but the principles all transfer easily to new pipework under construction. Note two important points about API RP 578.

- It is intended primarily for the verification of *alloy materials*, rather than plain carbon steels. This is because PMI techniques can differentiate more reliably between carbon and alloy steels than between closely related grades of carbon steel. Also, incorrect substitution of alloy steel with plain carbon steel could result in significant in-service problems such as corrosion, brittle fracture or high-temperature creep.

- PMI test methods and equipment are described in API RP 578. These are improving very quickly, however, with the result that even the quicker, cheaper test techniques are becoming more effective and accurate at differentiating between similar types of material.

Neither of these points has a dramatic effect on the scope of suitable SIFE exam questions; they just influence the possible wording a little.

22.2 The content of API RP 578

API RP 578 is a short document (about 15 pages), most of which contain short, concise definitions, statements and bits of technical information on material verification. There are few numerical data or detailed technical descriptions or justifications. The content consists of three main areas.

- What are the *objectives* of the PMI activity?
- *How* is the verification achieved?
- *Test methods and evaluation*: how do they work?

These subjects reflect the content of most of the SIFE exam questions you can expect to see on this subject.

With such a short document, there is little alternative if you wish to gain familiarity with it than to read through the whole thing, noting the progression and technical breakdown of the various chapters. Note the concise descriptions of the different PMI methods and the emphasis on their effectiveness at differentiating carbon steels from alloy materials, and the way that various responsibilities are allocated between the inspector and owner/user.

ARE YOU SURE?

- That you have picked out of API RP 578 what the owner/user (rather than the inspector) is responsible for? (You might be surprised.)
- You know API RP 578's opinion on where material verification activities are the most important?

If so, attempt question set 22a.

SIFE question set 22a: API 578

Q1. API 578 Scope

API 578 specifically does not cover:

(a) Carbon steel components
(b) Alloy steel components
(c) Stainless steel components
(d) Non-ferrous components

Q2. API 578 Definition

A certified document that permits each component to be identified according to the original heat of material from which it was produced is:

(a) PWHT report
(b) Weld map
(c) PMI (positive material identification) report
(d) Mill test report

Q3. API 578 Definition

PMI is a physical test to confirm that a material is consistent with:

(a) Material that is specified where mechanical properties and corrosion resistance is not an issue
(b) Specified alloy material designated by the owner/user
(c) Materials listed in ASTM specifications
(d) Any materials specified by the inspector (AI or SI, as applicable)

Q4. API 578 Extent

A written material verification programme indicating the type of PMI testing for existing piping systems should be established by:

(a) Owner/user
(b) Inspector
(c) Design engineer
(d) QA/QC engineers

Q5. API 578 Alloy substitution in CS service

Carbon steel systems in HF service are:

(a) Improved by using alloy steels as they are less hardenable

(b) Improved by using alloy steels as they harden more easily
(c) At risk from being substituted by alloy steels as they are more hardenable
(d) At risk from being substituted by alloy steels because they are more ductile

Q6. API 578 New QA material verification programme

A new construction QA material verification programme is applicable to:

(a) Carbon steel piping in pressure-containing components only
(b) Alloy piping in pressure-containing components only
(c) Carbon or alloy steel in pressure-containing components only
(d) Carbon or alloy steels in all components

Q7. API 578 Roles and responsibilities

Roles and responsibilities during a material verification (PMI) programme:

(a) Must be led by the inspector
(b) Must include the inspector
(c) Should include the material manufacturer
(d) Should be defined and documented

Q8. API 578 Roles and responsibilities

Who has the responsibility to verify that alloy materials placed into service are as specified?

(a) The owner/user
(b) The inspector
(c) The owner, user or their designee
(d) All parties involved in the job

Q9. API 578 Verification test procedure review

When should an owner/user or their designee review and approve the adequacy of a material verification programme?

(a) Never, it is repetition of effort
(b) When PMI testing is done by the material supplier
(c) When the inspection plan says so
(d) Always, to be on the safe side

Q10. API 578 Mill test report

Mill test reports are:

(a) A substitute for PMI test if witnessed by a third party
(b) Validated before shipping
(c) Certified by the owner/user
(d) An important part of a QA programme

Chapter 23

Surface preparation

The API SIFE body of knowledge (BoK) includes a concise group of standards on surface cleaning and preparation. In addition there is one that covers the measurement of paint coatings once they have been applied, that is, dry film thickness (dft). All of these are published by the Society for Protective Coatings (SSPC). Those in the list covering surface preparation are predominantly about the different shotblasting grades. Although high-quality 'white' or 'near-white' surface preparation is needed prior to most painting or lining applications, there are lower acceptable levels of surface quality that can be used for less important applications.

23.1 The SSPC quality hierarchy

Figure 23.1 shows the hierarchy of surface quality levels described by the SSPC 'SP' set of standards. These are all separate documents of only a few pages each. Most of the content of each of the standards is much the same for all of them in the series – they only really differ in the technical description of the usual appearance of the prepared metal surface in each of the individual levels. The numbers of the standards are not exactly consecutive as you move down the hierarchy. There is probably no hidden reason for this, it is just the way the standards developed. Note also that one of the lower levels, SP-14, is not included in the API SIFE BoK. A couple of other ones are included, however (see the bottom of Figure 23.1). For SIFE examination purposes, the main details you need to know about the various levels of surface preparation are as follows.

White metal blasting SSPC (SA-5)

This is the best possible standard of surface blast preparation under the SSPC hierarchy of standards. It is equivalent to the SA-3 designation

FIG 23.1
The SSPC surface preparation 'hierarchy'

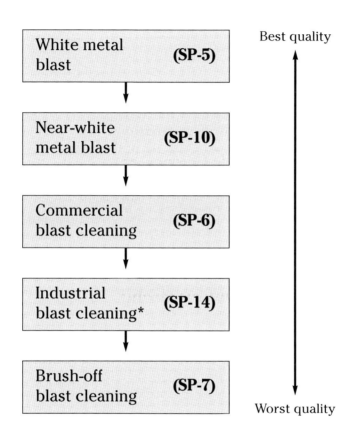

*SP-14 is not included in the API SIFE body of knowledge. A further four standards: Standard No.1 *Solvent Cleaning*, Standard No.2 *Power tool cleaning*, Standard No.11 *Power tool cleaning to bare metal* and paint application Standard No.2 *Dry coating thickness* are included.

used in Europe, from Swedish standard SIS 055900 (ISO 8501-1). It requires that *all* of the following contaminants must be removed by the blasting

- coating
- millscale
- rust
- oxides
- corrosion products
- any other foreign matter.

This is supplemented by the general requirement that visible deposits of oil and grease are removed before the blasting starts. Various different methods of blasting system and air circulation can be used, as long as the end result is achieved. In practice, an SA-5 blasted surface has a shiny, speckled appearance with *absolutely no staining*. This surface can be difficult and expensive to achieve – a contractor may have to change the blasting media several times to achieve the final unstained surface. Note again the content of the six points above; you will see the items again in the description of other levels in the SSPC surface preparation hierarchy.

Near-white metal blasting SSPC (SP-10)

This is the most common standard used as a minimum requirement for painting and lining of fixed equipment items. It is equivalent to the European designation $SA2\frac{1}{2}$ and represents a realistically achievable level for most blasting contractors without too much trouble. It differs from SP-5 in that, although most coating and millscale has still to be removed, a small amount of the surface (5%) is allowed to have residual *staining*. This staining is just discoloration, shadows and so on left over from the surface contaminants that have been blasted off. To the untrained eye or in poor lighting a near-white surface would look like a 'fully blasted' surface, perhaps with a slightly duller, less crystalline-looking surface.

Commercial blast cleaning SSPC (SA-6)

This has the same basis as near-white metal blasting SP-10 with the difference that up to 33% (rather than 5%) of the surface is allowed to show residual staining. It is common when the blasting media has become worn (and should really be replaced) or when time is limited. It can also be the result if a surface previously blasted to SP-5 or SP-10 has been left uncoated and open to atmosphere, when staining and subsequent corrosion will start within a few hours. SA-6 is a poor preparation to use as a basis for any permanent paint coating – you can expect some level of adhesion problems before too long.

Brush-off blast cleaning SSPC (SP-7)

Brush-off blast cleaning is a low level of preparation that does not remove tightly adhering surface coating such as millscale. It allows this to remain in position, but just roughens the surface. Millscale is defined as 'tightly adhering' if it cannot be lifted off with a thin-bladed knife. This SP-7 level is often used for maintenance-coating existing structural steelwork or when a less-than-perfect approach is adequate. It would not be suitable for new paintwork or linings, except in low-value utility steelwork applications such as shelters, doors, cladding and similar.

Power-tool cleaning to base metal: SSPC standard no. 11

This standard covers power tool cleaning rather than shotblasting. Power tool cleaning cannot clean pitted material very effectively, so it is acceptable for the surface to contain some rust or residual coating. This is therefore considered the power-tool equivalent of the standard for white metal SP-5, with the exception of the deposits that remain in the pitted surface. A blasting surface profile of 0.0001 inch (25 µm) is also required.

23.2 Dry coating film thickness (DFT) measurement conformance SSPC paint standard no. 2

This standard is the longest SSPC standard included in the API SIFE BoK. It consists of 12 pages, six of which are mandatory appendices covering dft measurement on girders, pipes, test panels and suchlike. The information in SSPC standard no. 2 is of good practical application during a source inspection of painting. As well as being a valid source of SIFE exam questions, it adds to the more generic information in Chapter 10 in Part A of this book.

What is difficult about paint dft measurement?

Nothing really. SSPC standard no. 2 explains the two types of thickness meter used

FIG 23.2

Dry film thickness (DFT) paint measurement

SOME KEY POINTS FROM SSPC PAINT STANDARD NO. 2

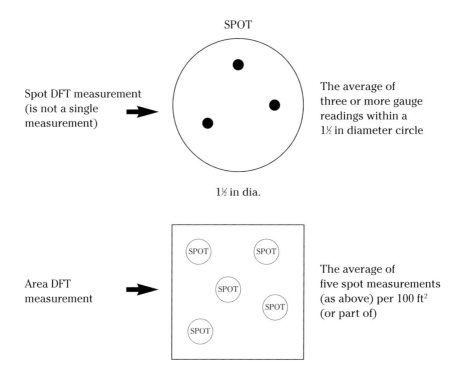

Spot DFT measurement (is not a single measurement) → The average of three or more gauge readings within a 1½ in diameter circle

Area DFT measurement → The average of five spot measurements (as above) per 100 ft² (or part of)

FOR LARGER COATED AREAS, USE THIS PROCEDURE

Area	Measurement
Up to 300 ft²	Select and measure each 100 ft² area
Between 300 ft² and 1000 ft²	Select and measure three 100 ft² areas
Exceeding 1000 ft²	Select one 100 ft² area in the first 1000 ft² plus one 100 ft² area for each additional 1000 ft² area (or part of)

Surface preparation

- magnetic *pull-off* types (type 1)
- electronic gauges using resistance or capacitance measurement (type 2).

These methods are described fully in the referenced standard ASTM D7091 if you want to know the full details.

Whereas the measurement of one individual dft point is easy, there are a few routines and procedures to be followed when measuring a component against an actual contract specification. Thickness readings have to be assessed against *spot* and *area* criteria, each with their own specific point and averaging measurement procedure. This is an essential (and real) activity for a source inspector (SI) and is not quite as simple as having a single dft figure that has to be met everywhere in a painted component. Figure 23.2 shows the procedures for spot and average dft measurement.

Dft acceptance criteria levels

Contract specifications for paint dft will usually specify a maximum and minimum acceptance thickness. There is provision in the standard for *five* acceptance (or restriction) levels around these requirements.

- **Level 1** is the most stringent and allows zero deviation in spot or area measurements from the specified maximum and minimum thickness.
- **Level 3** is the default level if it is not overridden by the contract specification. It allows extra tolerance to 80% of the minimum and 120% of the maximum specified dft thickness.
- **Level 5** is the most tolerant, allowing tolerances to 80% of the minimum specified dft thickness but no restriction on exceeding the maximum. In practice, this can result in paint over-thickness defects such as sagging, runs and flaking/cracking, particularly around edges and corners.

Quantifying non-conforming areas

If a coating is found to be non-conforming to the specified dft and acceptable level requirements then there is a specific procedure for quantifying what the degree of non-conformance actually is. This involves taking spot readings (as previously defined) at 5 ft (1.5 m) intervals in eight equally spaced directions radiating outwards from the non-conforming 100 ft^2 area. The objective is then to find (in each direction) a location that produces two consecutive non-conforming spot measurements. There would no doubt be many other measurement patterns that would give representative results, but you can treat this as a useful convention, if nothing else. As part of a published code it provides a clear convention to work to, and so can help avoid misinterpretation and arguments during source inspections.

Component-specific area dft measurements

Agreement between source inspectors and manufacturers can sometimes be difficult to achieve when area dft measurements of specific structural components are involved. Items such as girders must be coated with specific spray patterns to get the best results. I-beam girders in particular have 12 separate surfaces and lots of corners and edges, making it awkward to achieve uniform paint coverage. SSPC standard no. 2 gives guidance on these and other components. They are contained in the non-mandatory appendices, however, so are unlikely to be a common source for SIFE exam questions.

Exam questions on SSPC standard no. 2

Paint inspection and dft measurement is an important area for SIs, so it makes sense to expect some related questions to appear in the SIFE exam. Once again, the closed-book format of the exam places a restriction on the nature of the questions to be asked. Do not treat this standard too lightly, however; there are still some good bits of information in there that API could consider suitable for closed-book questions.

Before attempting the sample question set 23a below, read carefully through the SSPC standards, identifying the points mentioned from the start of this chapter. By now (if you have worked through this book correctly) you should be able to identify the types of facts and figures that fit the API exam style model of a valid closed-book question.

ARE YOU SURE?

- You know how spot and area dft measurements are taken?
- You have read the appendices of SSPC standard no. 2, in particular figure A1.2. (If you don't know what this is, then you haven't read it.)

If so, try the following question set 23a. You should be aiming to get 60%+ correct at your first attempt.

SIFE question set 23a: SSPC paint standard no. 2 (dft measurement)

Q1. Dft standards

The main standard for coating dft measurement techniques is:

(a) SSPC – SP2
(b) ASTM D7091
(c) SSPC paint standard no. 2
(d) SSPC standard no. 1

Q2. Dft standards

Coating dft conformance with specific contract requirements is covered by:

(a) SSPC paint standard no. 2
(b) ASTM D7091
(c) SSPC-SP2
(d) SSPC-paint standard no. 1

Q3. SSPC paint standard no. 2

SSPC paint standard no. 2 does not apply to:

(a) Epoxy coatings
(b) Thermal spray coatings
(c) Multi-coat paint coatings
(d) R primer-only paint coatings

Q4. Dft spot measurements SSPC standard no. 2

A dft spot measurement is:

(a) The average of three readings in a $1\frac{1}{2}$ in (40 mm) circle
(b) The average or lowest reading of three or more in a $1\frac{1}{2}$ in (40 mm) circle
(c) The lowest reading of at least three in a $1\frac{1}{2}$ in (40 mm) circle
(d) None of the above

Q5. Dft measurements: SSPC standard no. 2

When taking more than five readings within a spot then:

(a) The highest and lowest are disregarded

(b) Usually high or low readings that are not reported consistently are disregarded
(c) All the readings taken must be averaged
(d) The three lowest readings are the ones averaged

Q6. Dft area measurement: SSPC standard no. 2

The dft area measurement is defined as:

(a) The average of three or more readings optioned within a $1\frac{1}{2}$ in (40 mm) circle
(b) The average of the lowest five readings obtained over each 100 ft of surface
(c) The average of five spot measurements obtained over each 100 ft^2 of surface
(d) The full anathematic average of the readings obtained over a surface area of minimum 100 ft^2

Q7. Dft gauge types

The two main types of gauge for measuring dry film thickness are pull-off gauges and:

(a) Eddy current gauges
(b) Impression gauges
(c) Cross-hatching gauges
(d) Electronic gauges

Q8. SSPC standard no. 2: Dft measurement

When measuring coating dft, a true reading is obtained by comparing the spot or area readings with the base metal reading (BMR). The BMR is determined by:

(a) Averaging ten randomly spaced readings over the surface
(b) Using the lowest spot reading measureable in an uncoated area of base metal
(c) Averaging five readings within 1 ft (300 mm) of the centre of the surface
(d) Any of the above, depending what the component is

Q9. SSPC Standard no. 2: Area dft measurement

A coated area of $1200\,\text{ft}^2$ requires a minimum of how many individual dft readings to provide a representation area dft measurement?

(a) 5
(b) 10
(c) 15
(d) 20

Q10. Treatment of non-conforming dft measurements: Standard no. 2

If a dft for a $100\,\text{ft}^2$ area is found to be not a conformance with the contract specification then:

(a) Further thickness shall be applied
(b) The owner/user shall be informed
(c) A second, newly calibrated dft gauge shall be used to check the results
(d) Further readings shall be taken

Q11. Coating dft acceptance tolerances

A contract specifies a maximum dft of 500 microns with a level 5 tolerance restriction level. The maximum allowed spot reading anywhere on the component is:

(a) No limit
(b) 400 microns
(c) 600 microns
(d) 500 microns

Q12. Coating dft acceptance tolerances

The default level of dft measurement tolerance under SSPC standard no. 2 is:

(a) Level 1
(b) Level 3
(c) $+/- 10\%$ dft
(d) $+ 20\% / -10\%$ dft

Q13. Overcoating on low dft areas

If it is decided that an already coated surface requires overcoating then accurate dft measurements may be taken by using a Tooke gauge or:

(a) A type 2 gauge
(b) WFT readings
(c) A Buchholz gauge
(d) Any of the above

Q14. Coating of steel beam girders

In general, when measuring dft of an irregular component such as I-beam girders compared to a flat plate surface then

(a) Spot readings (as per SSPC standard no. 2) become less relevant
(b) Area readings become less relevant
(c) SSPC standard no. 2 is not suitable
(d) Additional spot readings are necessary per area

Q15. Dft gauge calibration: Standard no. 2

What is the calibration period for types 1 and 2 dft gauges specified by SSPC standard no. 2?

(a) There isn't one
(b) 12 months
(c) 12 months for type 1, 6 months for type 2
(d) 6 months for type 1 and 12 months for type 2

Q16. Dft measurement on coated pipes: standard no. 2

The number of dft measurements required to determine an 'area measurement' to SSPC standard no. 2 on externally coated pipework is:

(a) At least one per pipe spool
(b) The same as plates
(c) At least four per pipe spool (at cardinal points)
(d) SSPC standard no. 2 does not apply

Q17. Dft measurement of pipework components: standard no. 2

When measuring dft values for a wide selection of coated pipework spools accessories and other components loaded into a cart or rack in the works, compared to just plain pipe spools then:

(a) Dft measurements should be defined until arrival at the construction site
(b) WFT (comb gauge) measurements should be used instead
(c) More dft measurements are required
(d) Fewer dft measurements are required

Q18. Dft measurements rear edges

Dft measurements with a type 2 gauge of rear coated edges, for example, an I-beam girder require:

(a) Nominal spot readings as for a flat surface
(b) A minimum of three readings within $1\frac{1}{2}$ in of the edge
(c) Five readings spaced as close to the edge as possible
(d) Can be ignored as they will be inaccurate

Q19. Dft surface profile standards ref. standard no. 2

Surface profile grades are described in the standard:

(a) NACE 0175
(b) SSPC-SP10
(c) ISO 8503
(d) ISO 1940

Appendix 1

Answers to Part B sample questions

Question set 14a SI Responsibilities and process

Question	Answer	Ref (Study guide)
Qu 1	b	Ch1
Qu 2	a	Ch1
Qu 3	c	Ch1
Qu 4	c	Definition
Qu 5	a	Definition
Qu 6	b	Definition
Qu 7	c	Definition
Qu 8	d	Definition
Qu 9	b	6.1
Qu 10	d	7.1

Question set 14b SI Responsibilities and process

Question	Answer	Ref
Qu 1	c	8.2.1
Qu 2	d	8.2.7
Qu 3	b	8.2.8/8.3.2
Qu 4	d	8.3.3
Qu 5	a	8.4.2
Qu 6	c	8.5.4
Qu 7	b	8.6.3
Qu 8	c	General
Qu 9	b	8.6.4
Qu 10	c	8.7

Answers to Part B sample questions   327

Question set 14c SI Responsibilities and process

Qu 1	a	9.7.1
Qu 2	b	9.7.2.1
Qu 3	b	11.1.1
Qu 4	d	11.1.1
Qu 5	c	11.1.3
Qu 6	c	12.6
Qu 7	c	13.2.2
Qu 8	c	General
Qu 9	b	General
Qu 10	a	General

Question set 15a Material properties: API 577

Qu 1	b	10.3.1
Qu 2	c	10.3.2
Qu 3	d	10.3.5
Qu 4	a	10.3.6
Qu 5	b	10.4.1
Qu 6	c	10.4.2
Qu 7	c	General
Qu 8	a	10.4.2
Qu 9	c	10..4.3
Qu 10	b	10.4.3

Question set 15b Material properties: API 577

Qu 1	b	10.4.3
Qu 2	c	10.4.3
Qu 3	d	10.4.3
Qu 4	d	10.4.3
Qu 5	b	10.4.4
Qu 6	d	10.4.4
Qu 7	d	10.4.4
Qu 8	a	10.4.4
Qu 9	c	10.4.5
Qu 10	d	10.4.5

Question set 15c Material properties: API 577

Qu 1	c	10.4.4
Qu 2	c	10.4.5
Qu 3	b	10.4.5
Qu 4	c	10.4.5
Qu 5	d	10.5
Qu 6	b	10.5
Qu 7	b	General
Qu 8	a	General
Qu 9	d	10.55
Qu 10	d	General

Question set 16a NDE: API 577

Qu 1	c	9.8.9.4
Qu 2	a	9.11
Qu 3	a	9.7
Qu 4	c	9.8.9.4
Qu 5	b	Table 5
Qu 6	c	ASME V Art 6
Qu 7	b	9.3.1
Qu 8	d	9.2
Qu 9	a	9.3.2.3
Qu 10	b	9.4.1

Question set 16b NDE: API 577

Qu 1	b	9.10.2
Qu 2	b	9.1
Qu 3	c	9.4.2
Qu 4	a	9.5
Qu 5	a	9.5
Qu 6	c	9.6
Qu 7	b	9.6.1
Qu 8	b	9.11
Qu 9	b	9.7
Qu 10	d	9.7
Qu 11	d	9.4.1

Answers to Part B sample questions

Question set 17a Welding processes: API 577

Qu 1	a	5.2
Qu 2	b	5.1
Qu 3	b	5.3
Qu 4	a	5.3
Qu 5	d	5.4
Qu 6	c	5.6
Qu 7	c	5.6
Qu 8	b	5.3/5.7
Qu 9	b	5.3.1
Qu 10	b	5.4

Question set 17b NDE: Welding consumables: API 577

Qu 1	a	Sec 5
Qu 2	b	Sec 5
Qu 3	b	Sec 5
Qu 4	a	Sec 5
Qu 5	a	Sec 5
Qu 6	d	Sec 5
Qu 7	c	Sec 5
Qu 8	c	Sec 5
Qu 9	b	Sec 5
Qu 10	d	Sec 5

Question set 17c ASME IX welding qualifications

Qu 1	a	QW-101
Qu 2	d	QW-102
Qu 3	c	QW-102
Qu 4	b	QW-102
Qu 5	d	QW-251.1
Qu 6	d	QW-105.1
Qu 7	b	QW-105.1
Qu 8	c	QW-401.2
Qu 9	d	QW-401.1
Qu 10	c	Qw-401.3

Question set 18a AWS D1.1 Structural welding

Qu 1	c	General
Qu 2	c	Para 1.1
Qu 3	b	Table 6.1
Qu 4	a	Table 6.1
Qu 5	b	Fig 4.27

Question set 19a NDE: ASME B31.3 Ch 1

Qu 1	c	300.2e
Qu 2	d	300.2
Qu 3	b	Fig 328.5.2
Qu 4	b	300.2
Qu 5	a	300.2
Qu 6	a	300
Qu 7	c	300.2
Qu 8	b	300.2
Qu 9	d	300.2
Qu 10	b	300b

Question set 20a NDE: ASME VIII-I (UG)

Qu 1	a	UG-12/13
Qu 2	a	UG-77
Qu 3	b	Fig UG-84.1
Qu 4	b	UG-78
Qu 5	d	UG-81
Qu 6	a	UG-4f
Qu 7	b	UG-9
Qu 8	c	UG-10
Qu 9	b	UG-76
Qu 10	b	UG-4a
Qu 11	a	UG-77
Qu 12	c	UG-77c
Qu 13	a	UG-7
Qu 14	c	UG-80
Qu 15	a	UG-81

Answers to Part B sample questions   331

Question set 21a Valves/testing: API 598/S

Qu 1	b	API 598
Qu 2	a	API 598
Qu 3	b	API 598
Qu 4	c	API 598
Qu 5	b	API 598
Qu 6	d	API 598
Qu 7	c	API 598
Qu 8	a	Table 1
Qu 9	b	Table 2
Qu 10	a	API 598
Qu 11	a	Table 2
Qu 12	c	API 598
Qu 13	b	4.4.1
Qu 14	b	API 598
Qu 15	d	API 598

Question set 22a Material verification: API 578

Qu 1	a	Sec 1
Qu 2	d	Def 3.11
Qu 3	a	Def 3.13
Qu 4	a	4.1
Qu 5	c	4.1.1
Qu 6	b	4.2
Qu 7	d	4.2.1
Qu 8	a	4.2.1
Qu 9	b	4.4.2
Qu 10	d	4.2.4

Question set 23a SSPC standard no. 2 DFT

Qu 1	b	2.3
Qu 2	a	General
Qu 3	b	1.5
Qu 4	d	3.2
Qu 5	b	3.2
Qu 6	c	3.3
Qu 7	a	6.2
Qu 8	a	6.2

Qu 9	d	8.2.2
Qu 10	d	8.3
Qu 11	c	Table 1
Qu 12	a	Table 1
Qu 13	b	9.2
Qu 14	b	12.1
Qu 15	d	Annex 2
Qu 16	a	5.2
Qu 17	b	A7.1
Qu 18	c	A7.2
Qu 19	b	A6.3

Appendix 2

Steel terminology

Term	Definition
Alloy steel	This is a steel that has other (non-carbon) alloying elements such as manganese (Mn), nickel (Ni), chromium (Cr) and molybdenum (Mo) added to improve its mechanical properties (typically, strength, hardenability and corrosion resistance). Minimum amounts are required before it can be classified as 'alloy steel'.
Annealing	Heat treatment involving heat soaking followed by controlled cooling to soften and stress-relieve the steel.
Carbon equivalent (Ceq), given by $$\mathrm{Ceq} = \left[C + \left(\frac{Mn}{6}\right) + \left(\frac{Cr + Mo + V}{5}\right) + \left(\frac{Ni + Cu}{15}\right) \right]$$	This is perhaps the most important parameter in deciding how weldable a steel is. It uses the fact that other elements act like carbon, affecting a steel's tendency to crack during/after being welded.
Carbon steel (also plain carbon steel)	Steel with only residual amounts of elements other than carbon (e.g. Si, Al, Mg, Mn).
Chemical composition	This refers to the percentage of each element; iron (Fe), carbon (C) and all the other alloying and trace elements that are added for various improvement reasons. Each element has some effect, but chemical content alone does not match exactly with mechanical properties. The steel microstructure, heat treatment and form can have equally as much effect.
Chromium (Cr)	After carbon, chromium is the element that has the greatest effect on the hardenability of a steel. It also affects corrosion resistance, so its relationship with steel is an important one.

Term	Definition
Grains	As steel cools it orientates itself into a grain structure, which is visible under magnification. Grain structure is virtually unique to metals and sliding grains are what makes steel *ductile*, that is, able to plastically deform instead of cracking when under stress. Grain size has a major effect on how good a steel is at resisting crack propagation.
Hardenability	Steel that is hard is often quite strong (good), but may also be brittle, making it easier for cracks to propagate (bad). Material is hardened by heating and quenching (rapid cooling). How keen a steel is to be hardened is called its *hardenability*. It increases with %C content, and other elements – typically chromium.
Inclusions	Inclusions are 'foreign particles' such as oxides, or elements such as lead (Pb) or tungsten (W). They do not dissolve in the steel's structure, and so can act as defects, helping to initiate or propagate cracks.
Low-carbon steel	Steels with <0.15% C, used when maximum ductility is needed. Normally used in the hot-worked or annealed conditions. They do not have sufficient %C to benefit from the strength and hardenability that carbon provides.
Material *form*	The term *form* is used to describe the way in which steel is physically used to produce a manufactured component. The main ones are • pipes • tubes • plates • castings • forgings • wrought (mainly for fittings). Any of these may be made into assemblies by welding (fabrication).
Medium-carbon steel	Steel with 0.25–0.55%C. The higher carbon level makes it suitable for hardening and tempering to achieve strength and toughness.
Mild steel	Steel with <0.25%C. This starts to display increased strength as the carbon level gets above about 0.15%.
Normalising	Heat treatment by heating above a transformation temperature to relieve stresses and refine the grain structure.

Term	Definition
Phase	The same steel can exist in several different forms and patterns of atoms, molecules and crystals called *phases*. The simplest ones are *ferrite, austenite* and *martensite*, supplemented by many others. All have different mechanical properties that make the steel hard, tough, brittle and so on.
Steel	Steel is an iron-based alloy containing up to a maximum of about 2% carbon (C). Most weldable grades have <0.35%C.
Tempering	Reheating a hardened steel to reduce some of the hardness and increase toughness.
Weldability	This is the ability of a steel to be welded without suffering cracking during or after the welding process. As %C increases (above about 0.25%C), steel becomes less weldable, because carbon forms brittle compounds which just love to form cracks. An indication of weldability is given by the ASME 'P-number'. All low-carbon steels are P1, i.e. the least crackable – they are subdivided into four groups (groups 1, 2, 3, 4) based on progressive increase in their tensile strength.

Appendix 3

The ASME MDR form

The ASME MDR form

National Board Number: _____
Mfr. Representative: _____ Date: _____
Authorized Inspector: _____ Date: _____

FORM U-1 MANUFACTURER'S DATA REPORT FOR PRESSURE VESSELS
As Required by the Provisions of the ASME Boiler and Pressure Vessel Code Rules, Section VIII, Division 1

1. Manufactured and certified by _____
   (Name and address of Manufacturer)

2. Manufactured for _____
   (Name and address of Purchaser)

3. Location of installation _____
   (Name and address)

4. Type _____   _____   _____
   (Horizontal, vertical, or sphere)   (Tank, separator, jkt. vessel, heat exch., etc.)   (Manufacturer's serial number)

   _____   _____   _____   _____
   (CRN)   (Drawing number)   (National Board number)   (Year built)

5. ASME Code, Section VIII, Div. 1 _____   _____   _____
   (Edition and Addenda, if applicable (date))   (Code Case number)   (Special service per UG-120(d))

Items 6–11 incl. to be completed for single wall vessels, jackets of jacketed vessels, shell of heat exchangers, or chamber of multichamber vessels.

6. Shell: (a) Number of courses _____   (b) Overall length _____

| No. | Course(s) | | | Material | Thickness | | Long. Joint (Cat. A) | | | Circum. Joint (Cat. A, B & C) | | | Heat Treatment | |
	Diameter	Length		Spec./Grade or Type	Nom.	Corr.	Type	Full, Spot, None	Eff.	Type	Full, Spot, None	Eff.	Temp.	Time

Body Flanges on Shells											
									Bolting		
No.	Type	ID	OD	Flange Thk	Min Hub Thk	Material	How Attached	Location	Num & Size	Bolting Material	Washer (OD, ID, thk)	Washer Material

7. Heads: (a) _____   (b) _____
   (Material spec. number, grade or type) (H.T. — time and temp.)   (Material spec. number, grade or type) (H.T. — time and temp.)

| Location (Top, Bottom, Ends) | Thickness | | Radius | | Elliptical Ratio | Conical Apex Angle | Hemis. Radius | Flat Diameter | Side to Pressure | | Category A | | |
	Min.	Corr.	Crown	Knuckle					Convex	Concave	Type	Full, Spot, None	Eff.
(a)													
(b)													

Body Flanges on Heads											
									Bolting		
Location	Type	ID	OD	Flange Thk	Min Hub Thk	Material	How Attached	Num & Size	Bolting Material	Washer (OD, ID, thk)	Washer Material
(a)											
(b)											

8. Type of jacket _____ Jacket closure _____
   (Describe as ogee and weld, bar, etc.)
   If bar, give dimensions _____ If bolted, describe or sketch.

9. MAWP _____ _____ at max. temp. _____ _____ Min. design metal temp. _____ at _____
   (Internal) (External)   (Internal) (External)

10. Impact test _____ at test temperature of _____
    (Indicate yes or no and the component(s) impact tested)

11. Hydro., pneu., or comb. test pressure _____ Proof test _____

Items 12 and 13 to be completed for tube sections.

12. Tubesheet _____ _____ _____ _____ _____
    (Stationary (material spec. no.))   (Diameter (subject to press.))   (Nominal thickness)   (Corr. allow.)   (Attachment (welded or bolted))

    _____ _____ _____ _____ _____
    (Floating (material spec. no.))   (Diameter)   (Nominal thickness)   (Corr. allow.)   (Attachment)

13. Tubes _____ _____ _____ _____ _____
    (Material spec. no., grade or type)   (O.D.)   (Nominal thickness)   (Number)   (Type (straight or U))

(04/14)

National Board Number: _____

Mfr. Representative: _____ Date: _____

Authorized Inspector: _____ Date: _____

FORM U-1 (Cont'd)

Items 14–18 incl. to be completed for inner chambers of jacketed vessels or channels of heat exchangers.

14. Shell: (a) No. of course(s) _____ (b) Overall length _____

	Course(s)			Material		Thickness		Long. Joint (Cat. A)			Circum. Joint (Cat. A, B & C)			Heat Treatment	
No.	Diameter		Length	Spec./Grade or Type		Nom.	Corr.	Type	Full, Spot, None	Eff.	Type	Full, Spot, None	Eff.	Temp.	Time

								Body Flanges on Shells						
										Bolting				
No.	Type	ID	OD	Flange Thk	Min Hub Thk	Material	How Attached	Location	Num & Size	Bolting Material		Washer (OD, ID, thk)		Washer Material

15. Heads: (a) _____ (b) _____
    (Material spec. number, grade, or type) (H.T. — time and temp.)   (Material spec. number, grade, or type) (H.T. — time and temp.)

		Thickness		Radius		Elliptical Ratio	Conical Apex Angle	Hemis. Radius	Flat Diameter	Side to Pressure		Category A		
	Location (Top, Bottom, Ends)	Min.	Corr.	Crown	Knuckle					Convex	Concave	Type	Full, Spot, None	Eff.
(a)														
(b)														

						Body Flanges on Heads						
										Bolting		
	Location	Type	ID	OD	Flange Thk	Min Hub Thk	Material	How Attached	Num & Size	Bolting Material	Washer (OD, ID, thk)	Washer Material
(a)												
(b)												

16. MAWP _____ _____ at max. temp. _____ _____ Min. design metal temp. _____ at _____ .
    (Internal) (External)   (Internal)   (External)

17. Impact test _____ at test temperature of _____ .
    [Indicate yes or no and the component(s) impact tested]

18. Hydro., pneu., or comb. test pressure _____ Proof test _____

19. Nozzles, inspection, and safety valve openings:

Purpose (Inlet, Outlet, Drain, etc.)	No.	Diameter or Size	Type	Material		Nozzle Thickness		Reinforcement Material	Attachment Details		Location (Insp. Open.)
				Nozzle	Flange	Nom.	Corr.		Nozzle	Flange	

20. Supports: Skirt _____ Lugs _____ Legs _____ Others _____ Attached _____
    (Yes or no)   (Number)   (Number)   (Describe)   (Where and how)

21. Manufacturer's Partial Data Reports properly identified and signed by Commissioned Inspectors have been furnished for the following items of the report (list the name of part, item number, Manufacturer's name, and identifying number):

22. Remarks

(04/14)

The ASME MDR form

National Board Number: _____

Mfr. Representative: _____ Date: _____
Authorized Inspector: _____ Date: _____

FORM U-1 (Cont'd)

CERTIFICATE OF SHOP COMPLIANCE

We certify that the statements in this report are correct and that all details of design, material, construction, and workmanship of this vessel conform to the ASME BOILER AND PRESSURE VESSEL CODE, Section VIII, Division 1.

U Certificate of Authorization Number _____ Expires _____

Date _____ Name _____ Signed _____
                           (Manufacturer)   (Representative)

CERTIFICATE OF SHOP INSPECTION

I, the undersigned, holding a valid commission issued by the National Board of Boiler and Pressure Vessel Inspectors and employed by _____ of _____ have inspected the pressure vessel described in this Manufacturer's Data Report on _____, and state that, to the best of my knowledge and belief, the Manufacturer has constructed this pressure vessel in accordance with ASME BOILER AND PRESSURE VESSEL CODE, Section VIII, Division 1. By signing this certificate neither the Inspector nor his/her employer makes any warranty, expressed or implied, concerning the pressure vessel described in this Manufacturer's Data Report. Furthermore, neither the Inspector nor his/her employer shall be liable in any manner for any personal injury or property damage or a loss of any kind arising from or connected with this inspection.

Date _____ Signed _____ Commissions _____
                         (Authorized Inspector)   [National Board (incl. endorsements)]

CERTIFICATE OF FIELD ASSEMBLY COMPLIANCE

We certify that the statements in this report are correct and that the field assembly construction of all parts of this vessel conforms with the requirements of ASME BOILER AND PRESSURE VESSEL CODE, Section VIII, Division 1. U Certificate of Authorization Number _____ Expires _____.

Date _____ Name _____ Signed _____
                           (Assembler)   (Representative)

CERTIFICATE OF FIELD ASSEMBLY INSPECTION

I, the undersigned, holding a valid commission issued by the National Board of Boiler and Pressure Vessel Inspectors and employed by _____ of _____, have compared the statements in this Manufacturer's Data Report with the described pressure vessel and state that parts referred to as data items _____, not included in the certificate of shop inspection, have been inspected by me and to the best of my knowledge and belief, the Manufacturer has constructed and assembled this pressure vessel in accordance with the ASME BOILER AND PRESSURE VESSEL CODE, Section VIII, Division 1. The described vessel was inspected and subjected to a hydrostatic test of _____. By signing this certificate neither the Inspector nor his/her employer makes any warranty, expressed or implied, concerning the pressure vessel described in this Manufacturer's Data Report. Furthermore, neither the Inspector nor his/her employer shall be liable in any manner for any personal injury or property damage or a loss of any kind arising from or connected with this inspection.

Date _____ Signed _____ Commissions _____
                         (Authorized Inspector)   [National Board (incl. endorsements)]

(04/14)

Index

acceptance criteria 14
acceptance guarantees 38
allowable stress, S 36
API (ICP)
   API SIFE 'Code effectivity list' 193
   exams 191
   registering for the API SIFE
      exam 192–194
   SIFE programme scope 192
API exam questions 203–204
API SIFE exam questions 210
   some hidden secrets 195–203
   what to expect 195
ASME IX 259–260
   welding documentation formats 261
ASME MDR form 337–339
ASME VIII-1 288–300
   fabrication requirements 290
   inspection and testing 291
   marking and reports 291
      manufacturer's data report
         (MDR) 292
      nameplate and documentation
         requirements 292
   materials 288–291
   plate cutting, fitting alignment 294
   P-number (group) 295

Brinell hardness 58

carbon steels 44
   properties 45
code compliance 7
cultural differences 8

ductility 33, 35
dye penetrant testing (PT)
   inspection 82–85

FATT (fracture appearance
   transition temperature) test 38, 54–56
fillet weld 70
focus 24–26
full penetration weld 68, 69

hardness 33, 35
hardness tests 36, 56–58

impact test retests 59
impact tests 51, 53–54
inspecting materials 33–63
inspection co-ordinator role 214
ITP traceability requirements 62
ITP 13

levels of traceability: EN 10 204 47
linings inspection 170–186
   adhesion tests 179–180
   galvanising 183–186
      coating uniformity (Preece)
         test 184–186
      coating weight test 184
      hot dip galvanising 183
   ITPs for lined equipment 173–174
      rubber linings: ITP steps 174
   metallic linings 180–182
      loose cladding 181
   rubber hardness check 177–178
      International Rubber Hardness
         Degree (IRHD) scale 177
   rubber linings 170–172
      rubber-lined components' design
         features 172
   spark (holiday) testing 178–179
low-cost manufacture 8

magnetic particle testing (MT) 85–86
making decisions 28

malleability	35	preheat	71
manufacturer's data report (MDR)	14	pressure vessels, inspection of	116–148
ASME	142	ASME intent	123
material forms	38	ASME VIII intent	124
'material form' steel standards	39	dimensional check	135–139
material traceability	33, 46–50, 61	nozzle flange faces	138
material verification	40, 61	the tan line	138
positive material identification (PMI)	309	vessel 'bow' measurements	139
		vessel supports	139
material verification (API RP 578)	309–313	documentation review	127
metallurgy and materials: API RP 577	238–249	essential safety requirements (ESRs)	118
		European pressure equipment directive (PED)	118
non-compliance	7	head-to-shell misalignment	143
non-conformance reports (NCRs)	17	hold points	125
defending	22–24	hydraulic test	128
missing paperwork	21	inspection and test plans (ITPs)	125–127
tactics	19, 20	incomplete material traceability	143
non-destructive examination (NDE)	250–257	incomplete statutory certification	142
		incorrect weld preparations	144
painting inspection	155–169	manufacturer's data report	118
ASTM SSPC (Society for Protective Coatings)	162	maximum allowable working pressure (MAWP)	128
dry film thickness (dft)	163, 166–168	National Board	119
millscale	159	notified body (NoBo)	118
paint types	156–161	objectives of pressure test	127–128
air-drying	157	out-of-specification impact or hardness properties	145
primers	158		
two-pack paints	157–158	out-of-specification tensile properties	145
painting defects	165–166		
paintwork repairs	168–169	pneumatic testing	129–131
RAL numbers	163	pressure testing	127–133
reasons for paint failures	156	pressure vessel codes	147
surface preparation	158–159	statutory certification	117
shotblasting, points to check	160	unrecorded repairs	146–148
Swedish standard SIS 055900 (ISO 8501)	162	vacuum leak testing	131–133
		vacuum test procedure	132–133
pipework ASME B31.3	277–287	vessel markings	139–141
Cat D fluid	284	visual and dimensional examination	133–141
defect acceptance criteria	283		
heat treatment definitions	279	plate courses	134
pressure testing	283–284	procedure qualification records (PQRs)	64, 73–74
sensitive leak testing	283		
poor quality	7		
positive identification	34	radiographic testing (RT)	106–112
post-weld heat treatment (PWHT)	71	ASTM penetrameter (IQI)	112

sensitivity	109	mild steel	334
unsharpness	110	normalising	334
wire-type penetrameter	111	tempering	335
re-tests of material specimens	58–60	weldability	335
		strength	35
SIFE body of knowledge (BoK)	206	structural steelwork welding	270–276
SIFE exam preparation	225–230	ASTM SA-20: Steel plates for pressure vessels	273, 274
SIFE programme welding knowledge	258–259	ASTM SA-370: Material test methods	273, 274
SIFE study guide book	209–225	ASTM SA-6: Structural rolled steel	273
ASME and National Board (NB) code stamps	225, 226	AWS D1.1 Structural welding code	271
EPC	212	surface crack detection	81–88
high-risk manufacturing scenarios	218	dye penetrant testing (PT) inspection	82–85
indications, imperfections and defects	212	magnetic particle testing (MT)	85–86
inspection co-ordinator role	214	surface preparation	314–325
material test report (MTR)	214, 219	dry film thickness (dft)	314, 317–320
pre-inspection meeting	223	SSPC standard no. 2	317–319
rogue materials risk factors	219	Society for Protective Coatings (SSPC)	314
S/V	212	SSPC quality hierarchy	314–317
SI	212	brush-off blast cleaning (SP-7)	315
SIFE exam preparation	225–230	commercial blast cleaning (SP-6)	315
source inspection management programme	216	industrial blast cleaning (SP-14)	315
the inspection co-ordinator	212	near white metal blast (SP-10)	315
source inspection		white metal blasting SSPC (SA-5)	314–315
keeping your focus	27		
tactics	17		
why is it needed?	6		
source inspection process	231		
source inspector		technical argument	29
role	15	tensile strength	33
the basic skills set	25	tensile tests	36, 52
the management role	11	toughness	33, 35
Stahlschussel	40		
steel terminology	333–335	ultrasonic testing (UT)	89–94
chromium (Cr)	333	A-scan pulse-echo method	92
phase	335	of butt and nozzle welds	105
alloy steel	333	of castings	89
annealing	333	of forgings	89
carbon equivalent	333	of plate material	89
grains	334	of welds	96–97
hardenability	334	techniques	91–94
inclusions	334	EN 17640	98–104
low-carbon steel	334		
material form	334		
medium-carbon steel	334	valves and testing	301—308

API 598 Valve inspection and
    testing   301, 303
ASME B16.34 Flanged, threaded and
    welding end valves   301
    types of valve tests   302–304
        backseat test   302
        high-pressure closure test   302
        low-pressure closure test   302
        shell (or body) test   302
valves inspection   149–154
    ASME B16.34   149–154
        non-destructive examination
            (NDE) of valves   152
        valve classes   150–151
visual inspection of welds   79
volumetric non-destructive examination
    (NDE)   88–114

weld acceptance criteria   75
weld defects – how do they occur?   67
weld heat treatment   68
weld procedure specifications (WPS)
        64, 73
welder qualifications   74
welding
    acceptable level of defects   65
    checking weld preparations   77–79
    consumables   259
    defects   66–68
    documentation   71–76
    inspection   64–68
    ITPs   75–76
    processes: API 577 and ASME
        IX   258–269
welds, visual inspection   79